Joseph Reay Greene

A manual of the sub-kingdom Coelenterata

Second Edition

Joseph Reay Greene

A manual of the sub-kingdom Coelenterata
Second Edition

ISBN/EAN: 9783337175917

Printed in Europe, USA, Canada, Australia, Japan

Cover: Foto ©Andreas Hilbeck / pixelio.de

More available books at **www.hansebooks.com**

A MANUAL

OF THE

SUB-KINGDOM

CŒLENTERATA.

GALBRAITH & HAUGHTON'S SCIENTIFIC MANUALS

Experimental and Natural Science Series.

MANUAL

OF THE

ANIMAL KINGDOM.

II.
CŒLENTERATA.

BY PROFESSOR JOSEPH REAY GREENE.

SECOND EDITION.

LONDON:
LONGMAN, GREEN, LONGMAN, ROBERTS, & GREEN.
1863.

LONDON
PRINTED BY SPOTTISWOODE AND CO.
NEW-STREET SQUARE

A MANUAL

OF THE

SUB KINGDOM

CŒLENTERATA.

BY

JOSEPH REAY GREENE, B.A.

PROFESSOR OF NATURAL HISTORY IN THE QUEEN'S COLLEGE, CORK
&c. &c.

SECOND EDITION.

LONDON:
LONGMAN, GREEN, LONGMAN, ROBERTS, & GREEN.
1863.

PREFACE.

'THE house that is a-building looks not as the house that is built.' The present Manual, though now issued as complete, is, in truth, but the abridgment of part of a larger work which the Author trusts may yet one day see the light.

The general arrangement of the subject-matter here devised does not seem to require any explanation. Had the Author sought to evade those delays and difficulties with which, in almost every paragraph, he has found it his duty to contend, another, and far easier, plan might have been chosen. A thoroughly scientific method seemed, however, more likely to prove useful. And, in the discussion of questions hitherto considered unsusceptible of general treatment or, perhaps, insufficiently known to men of science themselves, he has not endeavoured, by the invention of difficulties which in nature have no existence, to hide truth beneath the patchwork veil of a meagre quasi-originality. Rather has it been his wish to unfold, in the clearest and simplest language at his command, phenomena which, as a student, he has himself earnestly striven to comprehend. Keenly, indeed, does he regret the deficiencies of style and want of artistic combination which, but too frequently, it is feared, will be found to mar his pages; believing, as he does, that for the interpreter of nature there is a standard of literary excellence not less high than that of the poet or historian.

In the select bibliographical list appended to the end of the Manual will be found the names of those writers from whose published works has been derived that assistance which the Author would now, gratefully, acknowledge. In particular to Professor Huxley are his best thanks due, for, without access to the original memoirs of that naturalist, the second chapter, on the Class *Hydrozoa*, could never have been rightly completed. But the Author must confess himself under deeper and less formal obligations to the same philosophic investigator, whose rich and suggestive seeds of thought could not, from their nature, fail to fall fruitfull on the soil of any patient mind.

From Professor Allman, also, who has done so much to promote a right knowledge of the *Cœlenterata*, the Author has not been denied kind aid.

And of foreign naturalists, personally unknown to him, he would especially single out, for courteous thanks, Professors Gegenbaur, R. Leuckart, Milne Edwards, and Agassiz.

To Mr. Gosse the Author is indebted for the loan of the beautiful drawings from which two of the woodcuts have been copied. The wood-engraver, Mr. William Oldham of Dublin, has executed his share of the following pages in by no means an unsatisfactory manner.

Mr. Busk, the Rev. Thomas Hincks, and Dr. Strethill Wright have also supplied the Author with some valuable facts touching the structure of the fixed *Hydrozoa*.

Queen's College, Cork:
April, 1861.

August, 1861.

Professor Max Schultze has just published a memoir on *Hyalonema* in which he confirms the opinion that the beautiful siliceous fibres of this organism are, in truth, to be regarded as spicules of a Sponge, allied, in some respects, to *Euplectella*.

Very recently, Professor Agassiz (op. cit. (71) p. 256), from personal examination of the living animal of *Millepora*, has concluded that the entire division of *Tabulata*, and, perhaps also, the *Rugosa*, can no longer be associated with the undoubted Actinoid polypes, but find, rather, their true place in the neighbourhood of the genus *Hydractinia*. The details of these observations having not yet fully appeared, it seems premature to adopt the important systematic change thereby indicated.

CONTENTS.

CHAPTER I.

THE SUB-KINGDOM CŒLENTERATA.

                                                                           Page

1. General characters.—2. Classes   •   •   •   3

CHAPTER II.

THE CLASS HYDROZOA.

SECTION I.

MORPHOLOGY AND PHYSIOLOGY OF HYDROZOA.

1. Type of the Class: Hydra.—2. General Morphology.—3. Organs of Nutrition.—4. Prehensile apparatus.—5. Tegumentary Organs.—6. Muscular System and Organs of Locomotion.—7. Nervous System and Organs of Sense.—8. Reproductive Organs   • • • 20

SECTION II.

DEVELOPMENT OF HYDROZOA • • • • 51

SECTION III.

CLASSIFICATION OF HYDROZOA.

1. Classification.—2. Order 1: Hydridæ.—3. Order 2: Corynidæ.—4. Order 3: Sertularidæ.—5. Order 4: Calycophoridæ.—6. Order 5: Physophoridæ.—7. Order 6: Medusidæ.—8. Order 7: Lucernaridæ • • • • 79

SECTION IV.

DISTRIBUTION OF HYDROZOA.

1. Relations to Physical Elements.— 2. Bathymetrical Distribution.—3. Geographical Distribution . . . 126

SECTION V.

RELATIONS OF HYDROZOA TO TIME 130

CHAPTER III.

THE CLASS ACTINOZOA.

SECTION I.

MORPHOLOGY AND PHYSIOLOGY OF ACTINOZOA.

1. Type of the Class: Actinia. — 2. General Morphology.— 3. Organs of Nutrition.—4. Prehensile apparatus.—5. Tegumentary Organs.—6. Corallum or Skeleton.—7. Muscular System and Organs of Locomotion. — 8. Nervous System and Organs of Sense.—9. Reproductive Organs   131

SECTION II.

DEVELOPMENT OF ACTINOZOA 170

SECTION III.

CLASSIFICATION OF ACTINOZOA.

1. Classification.—2. Order 1: Zoantharia.—3. Order 2: Alcyonaria.—4. Order 3: Rugosa.—5. Order 4: Ctenophora . 196

SECTION IV.

DISTRIBUTION OF ACTINOZOA.

1. Relations to Physical Elements. — 2. Bathymetrical Distribution.—3. Geographical Distribution . . . 231

SECTION V.

RELATIONS OF ACTINOZOA TO TIME.

                                                                            Page

1. General History of Actinozoa. — 2. History of Zoantharia.— 3. History of Rugosa.—4. History of Alcyonaria.—5. Silurian Corals.—6. Devonian Corals.—7. Carboniferous Corals. — 8. Permian Corals. — 9. Triassic Corals. — 10. Jurassic Corals. — 11. Cretaceous Corals. — 12. Tertiary Corals. — 13. Recent Actinozoa 236

BIBLIOGRAPHY OF THE CŒLENTERATA   .   .   . 249
QUESTIONS ON THE CŒLENTERATA   257
LIST OF ILLUSTRATIONS 261
INDEX   263

THE SUB-KINGDOM CŒLENTERATA.

CŒLENTERATA.

CHAPTER I.

THE SUB-KINGDOM CŒLENTERATA.

1. General characters. — 2. Classes.

1. General characters.—The animal forms included under the sub-kingdom *Cœlenterata* present modifications of a type of structure better marked than that which is characteristic of the *Protozoa*. All are furnished with an alimentary canal, freely communicating with the general, or somatic, cavity. The substance of the body consists essentially of two separate layers, an outer, or 'ectoderm,' and an inner, or 'endoderm.' These two membranes, but especially the former, are in general provided with cilia.

Another distinctive characteristic of the *Cœlenterata* is found in the presence of the peculiar urticating organs, or 'thread-cells,' which are met with so constantly in the integument of these organisms (*fig.* 1).

Thread-cells, for which the term 'cnidæ' has been proposed, usually occur as colourless, transparent, elastic, double-walled sacs, rounded or oval in form, and containing a fluid in their interior. The outer wall of the sac is entire and very delicate; the inner one is much stronger, having its open extremity produced into a stout, rather fusiform, sheath, which terminates in a long thread,

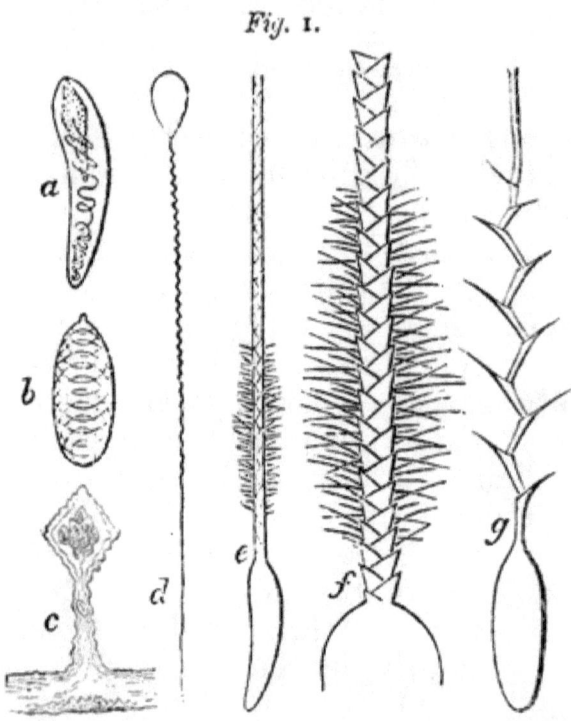

Fig. 1.

Urticating organs of COELENTERATA:—*a*, *e*, and *f*, thread-cells of *Caryophyllia Smithii*; *b*, thread-cell of *Corynactis Allmani*; *c*, portion of the marginal canal of *Willsia stellata*, with peculiar receptacle, containing thread-cells, arising therefrom; *d*, a single thread-cell of the same; *g*, thread-cell of *Actinia* (or *Bunodes*) *crassicornis*. (All magnified.)

or 'ecthorœum.' A number of barbs or hooks are sometimes disposed spirally around the sheath, the ecthorœum itself being often delicately serrated. In the ordinary condition of the thread-cell the ecthoræum lies twisted in many irregular coils round its sheath; the barbs of the latter being closely appressed to its sides, while it completely fills up the open end of the inner sac, into whose interior it projects. Under pressure or irritation, the cnida suddenly breaks, its fluid escapes, and the delicate thread is projected, still remaining attached to the sheath. So quickly is this done that the eye can by no means follow the process, but, in all probability, a complete eversion of the cell's contents takes place. In some cnidæ the presence of a sheath has not yet been discovered.

Thread-cells vary much both in form and size. They are unusually large in the Portuguese Man-of-war (*Physalia*), where they are spherical in figure and attain a diameter of ·003 of an inch. The relative dimensions of the thread and cell also vary. Sometimes the ecthoræum is scarcely longer than the sac; in other cases its length is nearly fifty times as great.

The disagreeable stinging sensations experienced when the human skin is brought into contact with the bodies of some *Cœlenterata* is, by most zoölogists, attributed to the influence of the thread-cells. It is supposed that the irritation is in part mechanical, arising from the friction of the filament or its sheath, and in part chemical, from the assumed poisonous nature of the fluid contained within the cell. The ease with which many Cœlenterate animals seize and, as it were, paralyze their struggling prey, is also ascribed to the same agency. These stinging propensities were evidently known to Aristotle, who refers to different forms of the present group under the name of ἀκαληφη, a term understood by some modern naturalists in a more restricted signification.

A few of the *Cœlenterata* are microscopic, but by far the majority are of appreciable size, and some attain considerable dimensions. Multiplication by gemmation is of common occurrence among the members of this sub-kingdom, and when, as frequently happens, the growths thus formed remain permanently in connection with the organism from which they originally sprouted, it is evident that this process, repeated several times, may give rise to aggregate masses, the limits of which it is not possible to define. In form the *Cœlenterata* vary considerably, presenting, in many cases, an external resemblance, sufficiently remarkable, to certain members of the vegetable kingdom.

The *Cœlenterata* possess no proper blood-vascular apparatus, distinct from the somatic cavity or its processes. The cilia which line the endoderm promote by their motion the circulation of the nutritive, or somatic, fluid occupying the general cavity of the body, and, in like manner, respiration is effected by the cilia of the ectoderm. Both of these ciliary movements are assisted by the contractions of the body walls, within which muscular fibres may, not unfrequently, be observed. Indications of a nervous system and organs of sense have been met with only in a few instances. Other structures, whose function would seem to be secretive, are not, however, wanting. Most of the *Cœlenterata* are provided with prehensile appendages, or 'tentacula,' and, in many of these animals, special organs, adapted for locomotion, are superadded. Throughout the entire of this department the elements necessary for discharging the function of true reproduction would appear to be present.

The power of emitting a phosphorescent light is eminently possessed by several *Cœlenterata*. This is more especially seen among the oceanic species which, together with *Noctiluca*, and other floating organisms, serve to produce the luminosity of the sea.

The Cœlenterate organism, therefore, has not only a plan of structure, or relative position of parts, peculiar to itself, but, viewed also as a mere animal machine, is seen to be, physiologically, in advance of the Protozoön. A comparison of the ultimate morphology of the two groups may serve further to elucidate this proposition.

The body of the Protozoön, as elsewhere we have shown, consists chiefly of the elementary tissue known as sarcode, or animal protoplasm; a soft, often transparent, elastic and extensile substance, albuminous in composition, and presenting the faintest traces of organisation.

The sarcode body is also remarkable for the manifold diversities of outward form which it may assume, though in many *Protozoa* there is little which deserves the name of integument, and an inner cavity, whether it exists under the form of contractile vesicle or alimentary track, is rudimentary in the highest degree. Some authors consider the Sponges as Cœlenterate, but the aquiferous system of these animals, however otherwise it may appear, is, in truth, lined by the outer surface of the organism.

Nevertheless, the homogeneity of the primitively simple sarcode is liable to become diversified by the two processes known as 'vacuolation' and 'fibrillation.' By vacuolation, clear spaces and granules arise in its substance, of which examples are furnished by *Actinophrys* and the *Gregarinæ;* by fibrillation, the same tissue may dispose itself in definite lines, as in the so-called stem muscle of *Vorticella*, and perhaps also the cortical investment of *Tethya*. Other structures, still further differentiated, are also seen to occur, as the nucleus, pigment-masses, reproductive elements, and the various kinds of cellæform bodies. But no true nervous or muscular tissues are produced, although these creatures manifest, in an humble manner it is true, some amount of contractility and sensibility.

An attempt might even be made to arrange the several forms of *Protozoa* in an artificial ascending series, the successive steps of which would differ in the relative degree of distinctness between the body-substance proper, and the outer portion, or conventional integument, to which it may give rise. In the lowest members of the group, as Dujardin has remarked, this external investment resembles nothing so much as the film which forms on the surface of flour paste when left to cool. Here pseudopodia are readily emitted from all parts of the body; but in *Pamphagus*, which, like *Amœba*, is naked, they are protrusible from one extremity only, the general surface of the body acting, as it were, the part of a more consistent membrane. From *Pamphagus* to *Difflugia*, and thence to the higher *Rhizopoda*, in which the outlying portions of the sarcode curiously differ, in their greater mobility, want of colour, and feebler tendency to undergo histological change, from the more highly vitalised body-mass within, it were not difficult to effect a natural transition. In the Sponges structural relations akin to those just mentioned are still more easily to be traced. In the *Gregarinæ* an external envelope becomes sufficiently distinct from the granular or vacuolated protoplasm which it bounds. And in the *Infusoria* the more contractile body-substance not merely serves to enclose a softer sarcode, but is itself protected by a cuticular covering, on which the styles and cilia are borne.

But the changes which the sarcode substance undergoes are not simply structural or mechanical. Other modifications, of a more purely chemical nature, may either accompany or replace the processes above mentioned. Thus, by 'conversion' into horny matter, the fibrous skeleton of the Sponges, the manducatory apparatus of *Chilodon* and its allies, the carapace of the *Arcellina* and *Infusoria*, and perhaps even their cilia, appear to be produced; or, by 'deposition' of mineral particles, withdrawn from the environment, shells and other hard structures have their origin. Nay more, the diverse forms of *Protozoa* have the power of appropriating certain elementary matters to the exclusion of others. The Polycystine sculptures its own siliceous shell; the Foraminifer, living beside it, a calcareous one, not less complex or beautiful; while from various parts of the body of the same Sponge a corresponding diversity of curiously wrought spicules may be obtained.

Thus, the naturalist, first struck by the variations in external aspect presented by the *Protozoa*, does not find his astonishment lessen when he begins to contemplate the manifold endowments of which each definable form is, as it were, the index, and perceives the fundamental sameness of organisation on which these complexities are based. All this, however, reflection should have led him to expect. For the body of every Vertebrate animal was once of as simple a structure as that of the Protozoön, and might even be said to correspond with it. But the life-history of the former plainly shows what it is capable of becoming. Some would add that the vital nature of the two organisms is less dissimilar than morphology and development would seem to indicate, and that those energies, which the lower animals spend so rapidly in acquiring the many outward modifications by which they are soon distinguished, might, if duly husbanded, and turned in another direction, give rise to very different structural products. Such a speculation is not wholly unworthy of mention. At present its discussion would lead us too far into the wide region of conjecture.

Turning now to the sub-kingdom *Coelenterata*, the members of this group are at once seen to differ widely from the *Protozoa*, in that their body-substance resolves itself into the two layers already mentioned under the names of ectoderm and endoderm; the former serving the purpose of an integument, the latter lining the large internal cavity constantly present.

These layers are very similar, though not identical, in structure. Both consist of a number of vesicular bodies, or 'endoplasts,' embedded in a homogeneous matrix, or 'periplast.' The endoplasts of the inner layer are more closely applied to one another, their size is somewhat larger, and their contents more transparent, than are the same parts of the outer layer. The chief difference between the layers is, however, in mode of increase, the ectoderm growing from within outwards, the endoderm from without inwards.

Even in *Hydra*, the lowest of Cœlenterate organisms, these two primitive layers are readily observable. But this animal exhibits, at certain seasons of the year, a tendency to break up into particles of a sarcode aspect, which retain for a long time an independent vitality. Nor are such amœboid masses wanting in the tissues of higher *Cœlenterata*. The significance of such facts should not escape our notice, since, at least, they serve to indicate the nature of the foundations on which the house of life has been constructed.

But the body-substance of the *Cœlenterata* by no means always presents that simple structure of its layers which the above expressions might seem to imply. Vacuolation and fibrillation here likewise perform their part, and the latter metamorphosis is often carried to such an extent as to give rise to true muscular fibres, though these are not, as in the higher animals, accompanied, in most cases, by an evident nervous system. Thread-cells, already described, the body-layers elaborate, as also reproductive elements, pigment-masses, and those other granular structures, which seem adapted for secretion. By conversion and excretion, outer growths are formed which serve either for support, ornament, or protection; while, by deposition of calcareous salts, the beautiful internal skeletons known as "corals" become variously produced.

Lastly, a wonderful diversity of external processes, some from the ectoderm, either alone or in great part, others from the ectoderm and endoderm combined, are seen to arise; and these may subsequently multiply to an almost indefinite extent, or, even separating from the primal organism, enjoy a brief but independent existence.

The student who has perused the account now given of the general structure of the lower animals is warned to be on his guard against the errors which still but too widely prevail on this and other kindred branches of histology. These errors, for the most part, have their origin in the well-known cell-theory of Schleiden and Schwann: a theory which, when first announced by its distinguished promulgators, possessed, indeed, a high dignity and utility, but in the hands of inferior naturalists has tended not a little to check independent thought, and to render obscure much that, but for it, would have been intelligible. In particular the view that the cell-nuclei, or endoplasts, constitute special originating centres of vital activity, is worthy of all reprobation; contradicted, as it appears to be, by so many careful observations on the various modes of development. Cells indeed exist, but only as the differentiated products of a primitively homogeneous protoplasm, not as morphological or physiological entities. To borrow the fine metaphor of Professor Huxley (whose views on animal structure we have here more than once sought to interpret and extend), "they are no more the producers of the vital phenomena, than the shells scattered in orderly lines along the sea-beach are the instruments by which the gravitative force of the moon acts upon the ocean. Like these, the cells mark only where the vital tides have been, and how they have acted."

A general survey of the development of Cœlenterate animals is best deferred till definition has first been given of their classes. Here, as in other sub-kingdoms, too little attention has been paid to vital changes succeeding what is called the "adult" condition, more especially to those which immediately precede the death of the organism. Such phenomena, in the present instance, would possess a peculiar interest; for individuality, the distinguishing characteristic of living beings, is displayed by many *Cœlenterata* in so remarkable a manner, that the phraseology by which, in the case of the higher animals, we are wont to designate its manifestations, cannot to the former be applied without considerable qualification.

Excepting two fresh-water genera, all *Cœlenterata* are marine. Few, if any, seas appear to want these animals, the several forms of which, both fixed and oceanic, enjoy also a varied bathymetrical range. The coral-reefs, so widely spread throughout the tropics, and the floating banks of jelly-fishes, amid which ships have been known to sail for days, testify, in like manner, to the abundance of a group of beings, whose place in the general economy of nature is not less extensive than significant.

Equally numerous were the *Cœlenterata* at former periods of the earth's history, nor does their wide distribution in space fail to find its parallel in a long-enduring existence through time. In the lower Silurian rocks Cœlenterate remains occur, to reappear in every stratified deposit of importance intervening between that epoch and the present day.

2. **Classes.** — Two leading modifications of structure may be traced among Cœlenterate animals, which admit, therefore, of being arranged under two principal groups or classes, the *Hydrozoa* and the *Actinozoa*.

In the *Hydrozoa*, the wall of the digestive apparatus is not separated from that of the somatic cavity, and the reproductive organs are external.

In the *Actinozoa*, the wall of the digestive sac is separated from that of the somatic cavity by an intervening space, sub-divided into chambers by a series of vertical partitions, on the faces of which the reproductive organs are situated.

In order to understand these relations aright, it will be necessary to contemplate, from a general point of view, the development of the *Cœlenterata*.

The scanty knowledge which we possess of the life-history of the *Protozoa* has, in previous pages, been sufficiently pointed out. From their known simplicity of structure, it may, indeed, be conjectured, that the changes which they undergo during the course of development must be comparatively slight. As already shown, the very existence of reproductive elements has yet to be ascertained in by far the greater number of the members of this sub-kingdom.

In the other four departments, true reproduction, by contact of ova and spermatozoa, universally occurs. Within the nutrient mass, or 'yolk, composing the bulk of the ovum, may be perceived a small cavity, the 'germinal vesicle,' which, in its turn, contains a still smaller particle, the 'germinal dot' (*fig.* 2, *b*). In addition to these parts, many ova are provided with an outer envelope, known as the yolk-sac or 'vitelline membrane.' The spermatozoa vary much in form. More com-

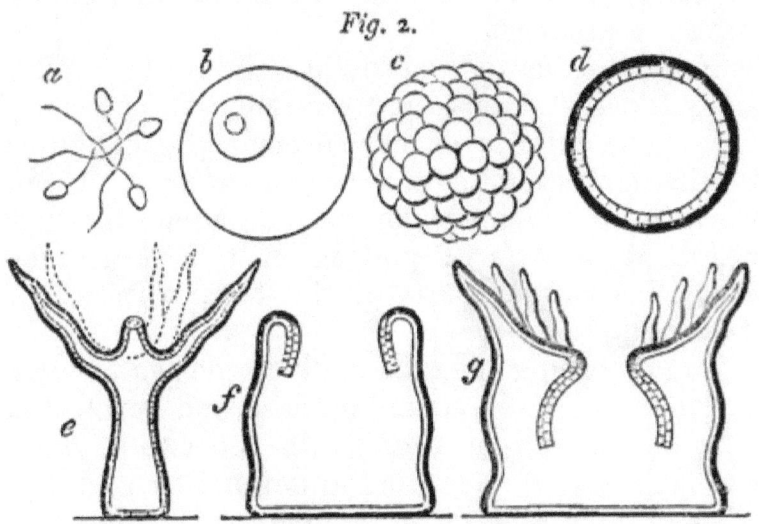

Fig. 2.

Development of Cœlenterata:—*a*, spermatozoa of *Cœlenterata*; *b*, section of Cœlenterate ovum, with germ-vesicle and germ-spot; *c*, ovum after segmentation; *d*, section of the same, more advanced, showing its division into two layers; *e*, longitudinal section of typical Hydrozoön; *f* and *g*, longitudinal sections of typical Actinozoön, in different stages of development. (These drawings are diagrammatic.)

monly they appear as delicate filaments, swollen at one extremity into a somewhat oval body (*a*).

After fecundation, the ovum exhibits a series of changes inaugurated by the process of 'segmentation' or yolk-division.

First, either the whole or a portion of the yolk

separates into two halves; each of these, again, into two others, and so on, until, finally, the segmented yolk is resolved into an aggregation of minute spherules, forming the so-called 'mulberry-mass' (c). The interior of this mass, in a large number of cases, liquefies, while its outer portion constitutes the 'blastoderm,' or 'germinal membrane,' from which the several embryonic structures are produced.

Next, the blastoderm divides into two layers, an outer, or 'serous,' and an inner, or 'mucous' (d). The outer layer more especially contributes to the formation of the organs of animal life, while from the inner layer is fashioned the first rudiment of a large portion of the alimentary canal, and various parts of the body with which it is connected.

The preceding remarks apply to the ova of most Cœlenterate, Molluscous, Annulose, and Vertebrate animals. In all of these, with the exception of the *Cœlenterata*, a further differentiation of the blastoderm would seem to take place. Each of its primary layers divides again into two others. In the outer portion of the serous layer arise the primitive rudiments of the nervous system, while the inner surface of the mucous layer forms the lining of the digestive canal. Between these two structures lies the " membrana intermedia " of embryologists, in the composition of which both serous and mucous layers appear, therefore, to take part. From this membrana intermedia the greater mass of the organic systems which make up the body of the adult animal are subsequently developed. Very soon after the formation of the intermediate layer, a number of important changes begin to ensue; but it is sufficient here to state, that, eventually, the principal nervous and vascular trunks are found to occupy opposite aspects of the body, the axis of which is traversed by the alimentary canal. Thus in every Vertebrate, Annulose, and Molluscous animal it is possible to recognise two distinct regions, a nervous, or 'neural,' and a vascular, or 'hæmal.'

In the *Cœlenterata*, on the other hand, no distinction between neural and hæmal regions can be noticed. Furthermore, whatever outward complexity the organism may present, all its parts are found on examination to be readily resolvable into the two layers previously referred to as ectoderm and endoderm. These correspond, both in structure and mode of growth, with the primitive layers of the germ in the higher animals, so that a general analogy may be traced between the permanent condition of the *Cœlenterata* and a well-marked embryonic stage in the *Mollusca*, *Annulosa*, and *Vertebrata*.

This is especially the case with the *Hydrozoa*, in which class a body-wall, composed of ectodermal and endodermal layers, invests the simple undivided cavity which constitutes the whole interior of the animal (*fig.* 2, *e*). The alimentary and somatic cavities are, therefore, in these beings identical, though, among many members of the group, certain parts of the organism are more immediately concerned in the performance of the digestive function.

But in the *Actinozoa* an oral fold of the blastoderm grows inwards to form a distinct digestive sac; thus, as it were, suspended, in the general cavity of the body, with which it communicates by means of a wide inferior aperture (*fig*, 2, *f* and *g*).

Some *Actinozoa* exhibit a further sub-division of each of the two primary blastodermal layers into other secondary membranes, foreshadowing a structure so constantly met with among the higher animals, just in the same manner as, in a few of the more advanced stomatode *Protozoa*, an indistinct differentiation of the body into layers indicates a condition which is manifested, without exception, by the immature forms of the four remaining sub-kingdoms.

DEVELOPMENT OF ANIMALS.

The organism does not exhibit true layers. PROTOZOA.	A blastoderm is formed, which divides into inner and outer layers.
The alimentary canal freely communicates with the somatic cavity. There is no distinction between neural and hæmal regions. COELENTERATA.	The two layers of the blastoderm become further differentiated. The alimentary canal has no direct communication with the somatic cavity. Distinct neural and hæmal regions appear.
The hæmal region is first developed. The mouth opens on the neural aspect. There is no segmentation of the blastoderm. MOLLUSCA.	The neural region is first developed. The blastoderm may divide into segments.
The mouth opens on the neural aspect, towards which the limbs are turned. ANNULOSA.	The mouth opens on the hæmal aspect, towards which the limbs are turned. A primitive groove, dorsal and visceral plates, are formed. VERTEBRATA.

Sub-kingdom CŒLENTERATA.

Animals whose alimentary canal freely communicates with the somatic cavity.

Substance of the body made up of two foundation membranes, an outer or ectoderm, and an inner or endoderm, which correspond, in mode of growth, with the primitive layers of the germ.

No distinct neural and hæmal regions. A nervous system absent in most.

Peculiar urticating organs, or thread-cells, usually present.

Class I. Hydrozoa.

Cœlenterata, in which the wall of the digestive sac is not separated from that of the somatic cavity, and the reproductive organs are external.

Class II. Actinozoa.

Cœlenterata, in which the wall of the digestive sac is separated from that of the somatic cavity by an intervening space, subdivided into chambers by a series of vertical partitions, on the faces of which the reproductive organs are developed.

CHAPTER II.

THE CLASS HYDROZOA.

SECTION I.

MORPHOLOGY AND PHYSIOLOGY OF HYDROZOA.

1. Type of the Class: Hydra.— 2. General Morphology.— 3. Organs of Nutrition. — 4. Prehensile apparatus. — 5. Tegumentary Organs.— 6. Muscular System and Organs of Locomotion.— 7. Nervous System and Organs of Sense. — 8. Reproductive Organs.

1. **Type of the class: Hydra.** — The *Hydra*, or fresh-water polype, is the type of the class *Hydrozoa* (*fig.* 2, *e*, and *fig.* 3).

The *Hydra* possesses a gelatinous, sub-cylindrical body, liable, from its contractility, to undergo various mutations of form, having one end expanded into an adherent disc, or foot, a mouth being situated at its opposite extremity. The mouth leads into a capacious cavity, excavated throughout the entire length of the animal. From the margin of the oral aperture, or rather, at a little distance below it, arises a circlet of prehensile tentacles, varying in number from five to twelve, or even more. These are exceedingly contractile, at one moment assuming the appearance of delicate filaments, the next, shrunk up into little wart-like knobs. Numerous thread-cells are embedded in their substance, and project freely from their surface. These are of two kinds, one being much larger than the other. The larger thread-cells, which are fewer in number, are further distinguished by the possession of a sheath, surrounded by three recurved barbs, and terminating in a long slender thread (*c*). The length of the *Hydra*, exclusive of its tentacles, seldom exceeds three fourths of an inch.

More minutely examined, the body of the *Hydra* is found to be composed of two membranes, an ectoderm and an endoderm. The former constitutes the outer layer of the animal, and has one of its sides always exposed to the water wherein the *Hydra* lives, the other side being in rather close contact with the endoderm, whose free surface forms the lining of the large internal cavity. The tentacles, which open into this cavity, are tubular prolongations of both the above membranes. It has been asserted by Trembley that the body of *Hydra* may be turned inside out, without thereby sustaining any injury, or being checked in the performance of its proper functions; but this experiment needs repetition. Both ectoderm and endoderm are vacuolated, especially the latter, and hence the well-known granular appearance which the *Hydra* presents under the microscope. Some describe the endodermal lining as produced into a number of villous elevations, projecting into the digestive cavity, and placed at right angles to its axis. The thread-cells are chiefly developed in connection with the ectoderm, and the numerous nodules, filled with these bodies, which the tentacles exhibit are merely enlargements of this outer layer. The ectoderm of the tentacles (and perhaps, also, of the body-wall) shows slight traces of the presence of muscular fibres. Around the margin of the mouth, the ectoderm and the endoderm unite together, and "the junction between the two is distinctly marked by a clear line."

The food of the *Hydra* consists, for the most part, of minute fishes, crustaceans, worms, and such other living creatures as come within the reach of its tentacles; and it is curious to observe, how, by means of these apparently fragile cords, animals are secured which would be deemed, at first sight, superior to their captor in strength and activity. There can, however, be little doubt, that the tentacles are aided in the performance of their prehensile function by the action of the thread-cells, with which they are so well provided. The elastic filaments of some of these are usually projected into the body of the captured organism, over whose motions they would appear to exert a potent benumbing influence, to the production of which the fluid contained in the interior of the cells probably serves to contribute. For it has even been observed that soft-bodied animals, which succeed in effecting their escape from the grasp of the *Hydra*, do not, in some instances, recover, but, soon afterwards, die. The entire surface of the tentacles is not at once brought into contact with the body of their victim, and, in the use of these organs, much instinctive caution is shown. When sufficiently mastered, the prey is thrust into the internal cavity, though the act of ingestion does not, in all cases, immediately put an end to its struggles. Gradually, the nutritive matters contained in its body are imbibed by the *Hydra*, all indigestible portions being finally expelled through the mouth. But some writers describe a short narrow canal, leading from the inner cavity to a small aperture, situate in the centre of the foot, through which particles of an excrementitious nature occasionally pass.

By means of its adherent disc the *Hydra* attaches itself to submerged stones, plants, and the surfaces of floating sticks or leaves. It is not, however, permanently fixed, but has the power of effecting changes in its position, either by the slow gliding motion of its base, or the repetition of certain leech-like movements, in which both tentacles and disc take part. Occasionally, the disc is protruded above the surface of the water, and, thus acting as a float, enables its possessor to remain, for a time, in this suspended condition.

The reproductive organs of the *Hydra* do not admit of being observed at all periods of the year, seldom making their appearance before the approach of the cold weather of autumn. Their position, however, is constant, the spermatozoa being contained in conical processes of the body-wall which arise close to the bases of the tentacles, while the ova are enclosed in rounded elevations of much greater size, and situated nearer the fixed extremity (*fig.* 3, *d*). Sometimes Hydræ are met with in which only one set of reproductive elements can be detected. Not more than one of the large protuberances, containing but a single ovum, is usually developed at the same time, and when two occur, they always arise from opposite sides of the animal. The ovum, having pushed itself through the body-wall, is seen to be invested with a spherical envelope, brownish or rosy in tint, and studded with a number of rough points, which some writers describe as bristles. Mr. Hancock, however, more properly regards them as

Fig. 3.

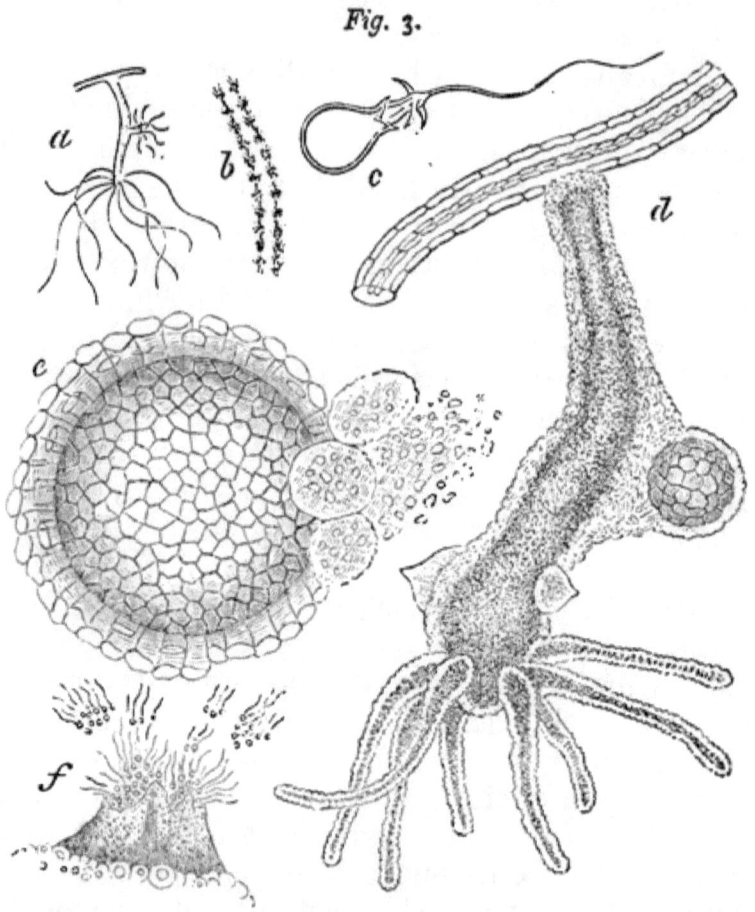

Morphology of HYDRA:—*a*, *Hydra vulgaris*, with a young polypite sprouting from its side, attached to a piece of stick; *b*, portion of a tentacle; *c*, one of its larger thread-cells; *d*, *H. viridis*, on a fragment of an aquatic plant, with elevations containing spermatozoa arising near the bases of the tentacles, between which and the attached extremity one side of the body-wall is seen to be much distended by an ovum; *e*, this ovum, ruptured; *f*, spermatozoa escaping from their receptacle, burst under pressure. (*a* is about twice the natural size; the others are much magnified.)

minute cells or sacs, "probably composed of some tenacious mucus with which to glue the egg to any substance on which it may happen to settle;" an inference supported by the fact, that, soon after the attachment of the egg, they disappear. Each spermatozoön consists of an oval body, furnished with a very delicate tail. The act of fecundation most probably takes place after expulsion of the ovum.

2. **General Morphology.**—Simple in structure as is the *Hydra*, it must, nevertheless, be viewed as the representative of an extensive assemblage of animal forms, whose outward aspect is singularly diversified, whilst, at the same time, the utmost uniformity prevails throughout all the more essential features of their organisation. In every Hydrozoön, the wall of the digestive apparatus coincides, or is continuous, with that of the somatic cavity, and the entire body, or 'hydrosoma,' howsoever modified, is resolvable into ectodermal and endodermal layers, processes of one or both of which constitute, in like manner, the manifold and complicated appendages, frequently superadded (*fig.* 5, *b*).

One end of the hydrosoma is always found to increase more quickly than the other, which in some cases, though not in all, is absolutely fixed. The term 'distal' is employed to distinguish this rapidly growing end from the opposite, or 'proximal,' extremity.

In *Hydra*, and a few of its more immediate allies, the hydrosoma consists, as has been shown, of an alimentary region, or 'polypite,' together

Fig. 4.

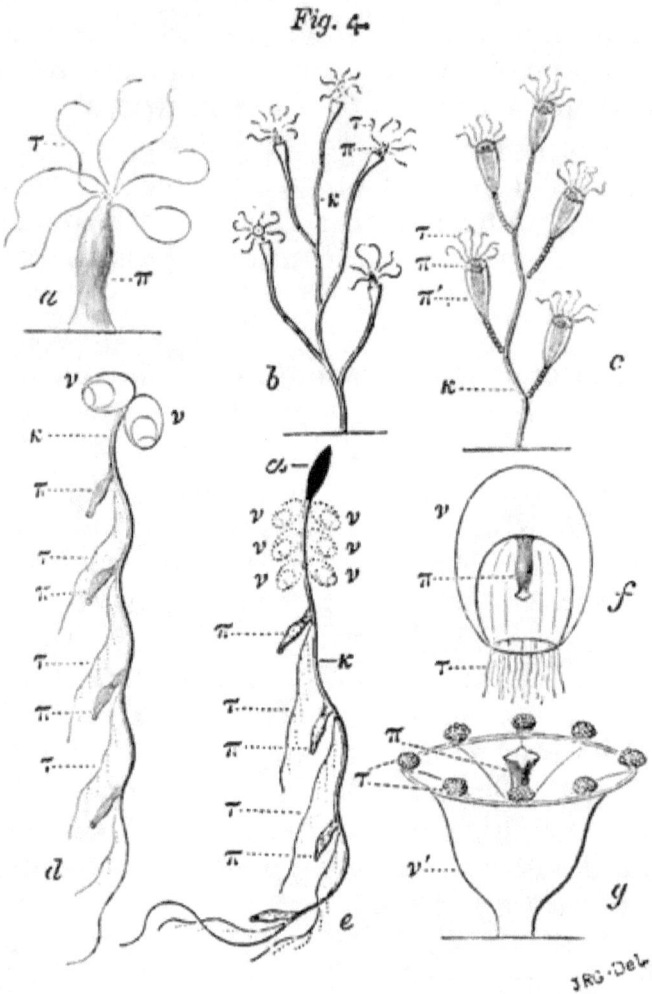

Morphology of Hydrozoa:—*a*, Hydrid; *b*, Corynid; *c*, Sertularid; *d*, Calycophorid; *e*, Physophorid; *f*, Medusid; *g*, Lucernarid;—π, polypite; τ, tentacles; κ, cœnosarc; π′, hydrotheca or polype-cell; ν, nectocalyx or swimming-bell; α, pneumatophore or float; ν′, umbrella. (These drawings are diagrammatic.)

with a 'hydrorhiza,' or adherent disc, prehensile 'tentacles,' and organs of reproduction (*fig. 4, a*).

More frequently, however, it is composed, not, as in *Hydra*, of a single polypite, but of several similar structures, connected with one another by means of a common trunk, or 'cœnosarc.' This cœnosarc may branch, presenting an erect tree-like aspect, and in such cases is permanently attached by means of the hydrorhiza which terminates its proximal extremity (*b*, and *fig.* 5, *a*). Often too, it excretes from its outer layer a strong chitinous investment, from which peculiar cup-shaped processes, or 'hydrothecæ,' serving as protective envelopes for the delicate polypites, may be developed (*fig.* 4, *c*). In other members of the class this firm layer has no existence, the cœnosarc remaining soft, flexible, and highly contractile; a modification which prevails among the complex oceanic *Hydrozoa*, creatures of great beauty and delicacy of structure, whose graceful movements through the element wherein they live are further assisted by the 'nectocalyces,' or swimming bells, with which the hydrosoma may be provided (*d* and *e*). In one group of these, the proximal end of the cœnosarc becomes transformed into a peculiar organ termed the 'somatocyst.' In others, this same extremity expands to form the 'pneumatophore,' or float, which enables its possessor to remain without effort near the surface of the water (*e*). There are, also, simple oceanic *Hydrozoa*, whose hydrosoma is represented by a single nectocalyx, from the under surface of which a polypite is, as it were, suspended (*f*). Finally, in another division of the group, the proximal extremity of the polypite is modified so as to form a special organ, the 'umbrella,' which usually simulates the function of a nectocalyx, though its structure and mode of development are very different (*g*).

In accordance with these several modifications, the class *Hydrozoa* has been divided into seven orders: *Hydridæ; Corynidæ; Sertularidæ; Calycophoridæ; Physophoridæ; Medusidæ;* and *Lucernaridæ.*

The *Hydrozoa* vary exceedingly in size. When a cœnosarc is present, it indicates a tendency on the part of the organism to increase by a process of continuous gemmation, and such forms often attain considerable dimensions, though the separate polypites are often so small as to be almost microscopic. This is the case with most of the complex fixed *Hydrozoa,* furnished with a branched horny cœnosarc. In the oceanic species, the polypites are somewhat larger, yet, when the hydrosoma consists of only one of these, its size is usually inconsiderable. But, should the hydrosoma develop an umbrella, its subsequent increase may be extremely rapid, and, in this manner, the most gigantic members of the class appear to be produced.

The animal fabric of the *Hydrozoa* is, perhaps, best described as consisting of —

    a. Organs of nutrition,
    b. Prehensile apparatus,
    c. Tegumentary organs,
    d. Muscular system and organs of locomotion,
    e. Nervous system and organs of sense, and
    f. Reproductive organs.

3. **Organs of Nutrition.**—In every Hydro-

zoön the entire somatic cavity may be said to perform the functions of a nutritive apparatus (*fig.* 5, *b*). But the true digestive process is chiefly effected within the bodies of the polypites.

Each polypite exhibits two regions, a distal, and a proximal. The distal extremity terminates in a delicate, more or less extensile, lip, which, in the *Calycophoridæ* and *Physophoridæ*, becomes everted and trumpet-shaped. Not unfrequently, the lip is lobed, its lobes, usually four in number, being, in some cases, very much prolonged (*fig.* 7, *b*).

In *Hydra*, and a few of the simpler forms of *Corynidæ*, the proximal end of the polypite is closed by the hydrorhiza, but throughout the remainder of the class, it freely opens into the somatic cavity.

Many *Calycophoridæ* and *Physophoridæ* have the proximal or attached division of the polypite produced into a more or less elongated peduncle, beyond which may be recognised two distinct regions; a median, or gastric, and a distal, or oral. The gastric cavity is separated from the interior of the proximal region by a peculiar inward growth, or 'pyloric valve,' which is best seen among the *Calycophoridæ*. Professor Huxley, its discoverer, describes it as "a strong circular fold of endoderm, whose lips, when the valve is shut, project into the cavity of the gastric, or median, division of the polypite. As the oily or albuminous globules which result from the digestive process are formed, they usually accumulate close to the valve, and are kept constantly rotating by the cilia which line the gastric chamber. After remaining for

Fig. 5.

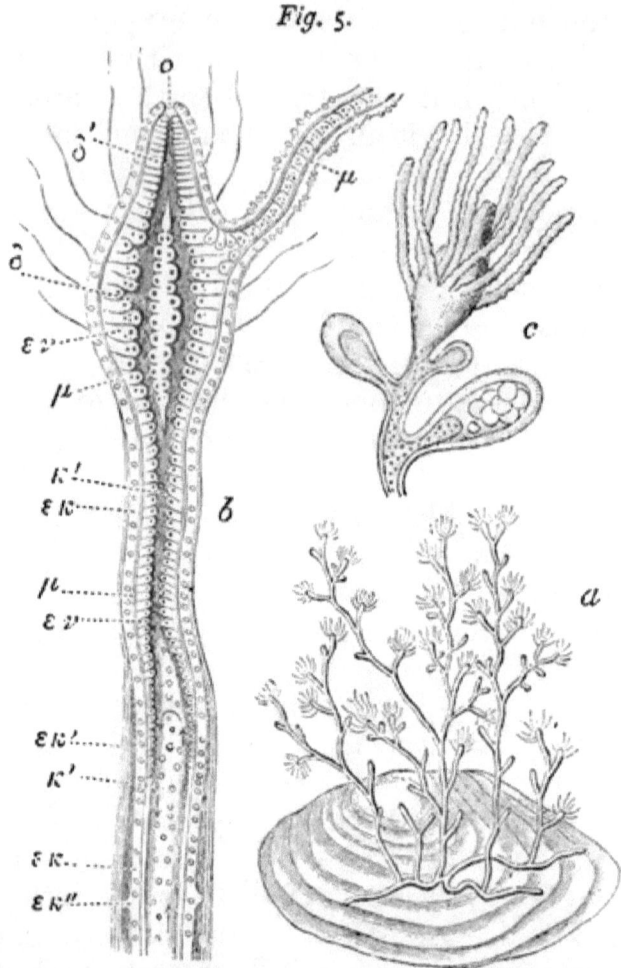

Morphology of CORDYLOPHORA:—*a*, hydrosoma of *Cordylophora lacustris*, growing on a dead valve of the swan-mussel; *b*, longitudinal section of a portion of its cœnosarc, with a single polypite; *c*, another fragment of the same cœnosarc, to which are attached, a polypite, and three reproductive bodies, or gonophores, in various stages of development, the largest and most advanced containing ova;—*o*, mouth; δ', cavity behind the mouth; δ, gastric cavity; κ', interior of the cœnosarc; $\epsilon\nu$, endoderm; μ, muscular layer; $\epsilon\kappa$, ectoderm; $\epsilon\kappa'$, outer hard layer, or polypary, excreted by ectoderm; $\epsilon\kappa''$, processes of ectoderm, connecting it with the inner surface of this hard layer. (All, except *a*, magnified.)

a while in this position, the fundus of the gastric chamber contracts, and forces the globule through the valve, which appears to dilate at the same moment."

The walls of the peduncle are thin and muscular. Those of the gastric chamber are comparatively thick, while its cavity is wider than any other part of the interior of the polypite. In addition to the cilia, which clothe its endoderm, the surface of this layer is often elevated into a number of villi, or conical processes, which in *Physalia* attain a length of ·01 of an inch. Such villi, or ridge-like enlargements which arise in their stead, have been observed within the polypites of many *Hydrozoa* (*fig.* 5, *b*). The coloured contents, occasionally noticed in these, and similar elevations, are regarded by some anatomists as the most rudimentary form of hepatic apparatus. In *Velella* and *Porpita*, more certain indications of a liver are presented by a dark brownish mass, which arises in connection with the digestive cavity of the large central polypite (*fig.* 21, *a*).

The nutrient matters elaborated within the bodies of the polypites are finally transmitted to the somatic cavity (*fig.* 5, *b*). Here they undergo an imperfect sort of circulation, a phenomenon the occurrence of which has been more especially observed within the long cœnosarc of *Tubularia indivisa*. Currents have been seen to course up and down the long stem of this Hydrozoön which occasionally appear to flow through distinct tubes, but these are nothing more than irregular cavities produced by vacuolation of the endoderm (*fig.* 16, *c*, *d*, and *e*). A circulation of albuminous particles also takes place within the peculiar nutrient system of the *Medusidæ* and *Lucernaridæ* (*fig.* 23).

The cavity of the cœnosarc, or of the peduncles of the polypites in connection therewith, may become, as in *Tubularia*, partially obliterated by vacuolation, a process, however, which does not seem to impair its vital efficiency.

Some have conjectured that the short canal which penetrates the attached extremity of the *Hydra* represents the cœnosarcal cavity of the higher *Hydrozoa*.

4. **Prehensile apparatus.**—The tentacles, or prehensile organs, which present so striking a feature in the physiognomy of the *Hydrozoa*, vary exceedingly, both in position and structure.

In the *Hydridæ*, *Sertularidæ*, and some genera of *Corynidæ*, they arise, in one or more circles, either immediately around the mouth or at a short distance below it (*fig.* 3, *d*). But, in other *Corynidæ*, they spring from various parts of the walls of the polypite (*figs.* 16 and 17), and one genus of this group, *Hydractinia*, possesses, in addition, long isolated tentacles, having an independent origin of their own from the cœnosarc. Single tentacles of this kind are the only ones which occur in the genera *Physalia*, *Velella*, and *Porpita*, among the *Physophoridæ* (*fig.* 11, *c*). In other *Physophoridæ*, and all *Calycophoridæ*, the tentacles are inserted on the sides of the polypites, about the junction of the gastric and proximal divisions (*figs.* 20 and 22). In the *Medusidæ* and *Lucernaridæ*, the tentacles usually surround the open border of the bell-shaped swimming organ (*figs.* 7, 23-25).

Transverse section of a tentacle shows it to consist of ectodermal and endodermal layers, enclosing a process of the somatic cavity, which, in very many cases, is wholly obliterated. The ectoderm is often found to exhibit muscular fibres, and is always provided with numerous thread-cells, which may accumulate near its surface in minute rounded papillæ, imparting to the tentacle a roughened appearance (*fig.* 19, *b*).

Except in the *Calycophoridæ* and *Physophoridæ*, the tentacles are very seldom branched. In some genera of these orders the tentacles are as simple in structure as those already described, but, in others, they attain a much greater complexity. Each tentacle of *Physalia* has a sac-like expansion at its base, while numerous reniform enlargements, each well packed with thread-cells, are disposed transversely along its sides (*fig.* 11, *d*). Both these and the sac communicate with a canal which runs through the entire length of the tentacle. The side opposite the reniform enlargements is bordered by a wide muscular band, which connects itself, above, with the basal dilatation. The reniform enlargements are, in other genera, replaced by lateral branches, some of which present three well-defined regions: a 'pedicle,' or proximal slender portion; a 'sacculus,' or "median division, with one wall much thicker than the other, containing numerous elongated thread-cells, arranged in transverse rows perpendicularly to the wall, and flanked on each side by a longitudinal series of larger oval thread-cells;" and, finally, a 'filament,' or "terminal cylindrical thread, full of large rounded thread-cells." Such, according to Huxley and Kölliker, is the structure of the tentacular branches in *Forskalia*, and which, under slight modifications, repeats itself in many other forms of *Physophoridæ* and *Calycophoridæ*. Within the sacculus of the last-mentioned order, occurs a peculiar zig-zag muscular cord, known as the "angel-band," the nature of which is but imperfectly known.

Where the pedicle and sacculus unite, a solid process of the ectoderm has been observed to originate in some *Physophoridæ*. From its investing the sacculus in the form of a hood, this organ has received the name of 'involucrum.'

The mode of action of the tentacles, as appendages for prehension, has been sufficiently explained in our account of the *Hydra*.

The name of 'nematophores' has been given by Mr. Busk to peculiar cæcal processes, distinct from the oral tentacles, which are found on the cœnosarc of some *Sertularidæ*. Like the rest of the cœnosarc these processes are invested with a stiff, horny, layer, open at the distal end of the nematophore, beneath which are embedded many large thread-cells. The nematophores probably serve as organs of offence. They are most numerous in the genus *Plumularia*.

5. **Tegumentary Organs**.—The tegumentary system in the *Hydrozoa* is composed wholly of the, in general, ciliated, ectoderm, and the rich supply of thread-cells to which this layer gives rise. Usually it appears more or less vacuolated, or it may even become changed into a gelatinous mass. The thickened disc of the *Medusidæ* and *Lucernaridæ*, in some structureless or but faintly

cellular, in others has its homogeneous periplast traversed in all directions by a complex mesh-work of threads, which remain quite distinct from the endoplasts about which they diverge, and with whose processes they appear never to become continuous. The threads themselves seem elastic, transparent, of different diameters, frequently dividing, soon to unite again, and, occasionally, disposing themselves side by side in such a manner as to form extended plates. On the convex aspect of the soft mass, which this thread system strengthens, the surface of the periplast is broken up into a number of polygonal cells, each furnished with an endoplast; and in the delicate epithelial layer thus produced thread-cells may, without difficulty, be observed.

In the *Corynidæ* and *Sertularidæ* the ectoderm excretes a firm, structureless, cuticular lamina, to which the name of 'polypary' has been restricted by Professor Allman. This may so far separate itself from the outer surface of the ectoderm as to present, at first sight, the aspect of a distinct layer (*fig.* 5, *b*,). In such cases its connection with the ectoderm is maintained by means of transverse processes arising from the latter, and these may present (*fig.* 19, *b*) considerable regularity of arrangement. The cup-shaped chambers, or hydrothecæ, commonly known as polype-cells, which are so characteristic of the order *Sertularidæ*, are merely prolongations of this excreted layer.

The polypites of the *Calycophoridæ* and *Physophoridæ* are, in some genera, protected by overlapping appendages, termed bracts, or 'hydrophyllia.' These derive their origin from both ectoderm and endoderm, though chiefly from the former, and always enclose a 'phyllocyst,' or cæcal process of the somatic cavity.

The pigment-granules of the *Hydrozoa* are, in general, of irregular form, and, though usually found in the ectoderm, are by no means, as we have seen, peculiar to it.

6. **Muscular System and Organs of Locomotion.**—In connection with the more or less contractile body-substance itself, separate muscular fibres, whose arrangement is most frequently longitudinal, present themselves among many *Hydrozoa*, and especially in the highly contractile cœnosarc of the *Calycophoridæ* and *Physophoridæ*. Similar fibres may also be traced in the walls of the polypites and tentacula, or on the concave surface of the swimming organs, with which several of the oceanic species are provided. They occur most abundantly in the ectodermal layer (*fig.* 5, *b*). In the *Medusidæ* and *Lucernariidæ* both radiating and circular fibres, not without distinct indications of transverse striation, have, by different observers, been detected.

Special swimming organs, or nectocalyces, are found in the *Calycophoridæ*, *Medusidæ*, and many of the *Physophoridæ* (*figs.* 20, 22, and 23). Each nectocalyx is a cup, or bell-shaped body, the open cavity of which is provided with a muscular lining, and has received the name of 'nectosac.' Around the margin of the nectosac, the wall of the nectocalyx is produced inwards, forming a shelf-like membrane, or 'veil.' By means of this membrane, which is highly contractile, the aperture of the bell may be more or less narrowed. Within the thickness of the roof of the nectocalyx there occurs, also, a second cavity, from which issue four or more canals, having no direct communication with the nectosac, beneath the walls of which they are prolonged until they reach a circular vessel surrounding its margin (*fig.* 23). The substance of the nectocalyx consists chiefly of ectoderm, but a continuation of the endoderm lines the 'nectocalycine canals' and the cavity from which they arise.

The function of the nectocalyx is sufficiently simple. By the rhythmic contractions of its muscular lining, the water within the nectosac is expelled, and the organism moves in a contrary direction.

The umbrella of the *Lucernaridæ* bears some resemblance to a nectocalyx, and, like it, may perform the function of a natatorial organ. It is easily distinguished by the absence of a veil: its size, also, in the great majority of cases, is much more considerable (*figs.* 7 and 25).

7. **Nervous System and Organs of Sense.** —The swimming organ, whether it take the form of a nectocalyx or umbrella, is usually furnished around its margin with a number of supposed organs of sense, known as the marginal bodies. For the best account yet given of their structure and relations, we are indebted to the researches of Will and Gegenbaur.

Two kinds of these bodies are found in the *Medusidæ*,—'vesicles,' and 'pigment-spots.'

The vesicles are thin-walled, rounded or ovate, sacs, lined internally with an epithelial layer, and containing one or more solid, motionless concretions, immersed in a transparent fluid. These concretions are oval or spherical in form, and appear to be composed of carbonate of lime. From the inner wall of the comparatively large vesicle in *Geryonia* arises a short stalk, which expands to form a delicate membrane around the solitary concretion. In some forms, a much thicker covering invests, along with the concretion and a number of minute molecules, a round or oval body, not unlike an endoplast. Many other modifications of the vesicles have been described.

The pigment-spots, otherwise termed "ocelli" and "eye-specks," consist of aggregations of colouring matter, enclosed in distinct cavities. The tint of these bodies is often extremely brilliant, shades of yellow, red, and black being most predominant.

Oceania turrita is the only known Medusid in which vesicles and pigment-spots co-exist (*fig.* 23).

In the umbrella of the *Lucernaridæ*, both vesicles and pigment-spots seem to become united into a single organ, termed the 'lithocyst.' These marginal bodies are protected, externally, by a sort of hood, and present often a very complex arrangement. Most frequently they occur as ovate vesicles, mounted on short, hollow stalks, each of which communicates by means of a canal with one of the radiating vessels of the umbrella. Within, the vesicle is delicately ciliated, and encloses at its free extremity a broad, thin-walled, oval sacculus, packed with a number of six-sided crystalline prisms, obliquely truncated at each end.

Many zoölogists describe the vesicles as "auditory organs," the crystals or other bodies which they contain being designated "otolites." But this view of their nature is altogether hypothetical. Gegenbaur hints that they, perhaps, constitute an apparatus of excretion. The pigment-spots may be regarded with some shade of probability as the earliest indications of organs of sight, which appear among the lower forms of the animal kingdom. In some ocelli, a spherical, highly refractive corpuscle has been detected by Gegenbaur within the mass of pigment.

It is by no means certain that a nervous system exists in any of the *Hydrozoa*. Professor Agassiz describes what he considers as such a system in the nectocalyces of some of the free swimming forms. "In Medusæ (he writes) the nervous system consists of a simple cord, of a string of ovate cells, forming a ring around the lower margin of the animal, extending from one eye-speck to the other, following the circular chymiferous tube, and also its vertical branches, round the upper portion of which they form another circle. The substance of this nervous system, however, is throughout cellular, and strictly so, and the cells are ovate. There is no appearance in any of its parts of true fibres." But the structure of the tissues here described as nervous is very susceptible of a different interpretation. M'Crady and Fritz Müller have also speculated on the presence of a nervous system in the *Medusidæ*. The former naturalist states that, among several species, he has observed a distinct ganglion in the neighbourhood of each marginal body.

Among the varied appendages attached to the cœnosarc of the *Physophoridæ* occur certain processes of doubtful nature, which Kölliker and some other observers appear disposed to regard as organs of touch (*fig.* 22). The 'hydrocysts,' or feelers, for so have these bodies been termed, bear some resemblance to polypites in their structure and mode of attachment, but differ from them in possessing cæcal extremities, beneath which large thread-cells are embedded. In some genera they are with difficulty distinguished from polypites in a young state of development.

8. **Reproductive Organs.**—Processes of the body-wall, within which are developed true generative organs, the 'spermaria' and 'ovaria,' constitute the reproductive apparatus of the *Hydrozoa*. These processes are always external, and are remarkable for the interesting series of modifications which they present among the several members of the class (*figs.* 6 and 7).

In *Hydra*, as already shown, they are of the simplest possible structure, differing from other parts of the body-wall in their contents alone. Such, also, is their aspect in *Lucernaria*, *Pelagia*, and, probably, all the *Medusidæ*.

In the *Corynidæ, Sertularidæ, Calycophoridæ,* and *Physophoridæ*, the reproductive bodies appear externally as distinct buds, or sacs, for which Professor Allman has proposed the name of 'gonophores.'

The simplest kind of gonophore consists of a well-defined protuberance from the body-wall, the 'sporosac,' containing within its substance ovaria or spermaria, and enclosing a diverticulum of the somatic cavity (*fig.* 6, *a*). The proximal extremity of this and other kinds of gonophore usually becomes narrowed into a short stalk of attachment. Such a form of the reproductive bud is of comparatively rare occurrence. Examples may, however, be found in some species of *Hydractinia*, *Coryne*, and *Clava*.

To understand the structural modifications of the gonophores in other *Hydrozoa*, it is necessary to trace briefly the principal phenomena which the more complex of these bodies present in the course of their development.

All gonophores first appear as simple processes of some portion of the body-wall, with its two layers, the ectoderm and endoderm. Next, the process becomes better defined, and exhibits a peduncle or stalk. " When this process has attained a certain size, its distal wall becomes thickened, and projects as a sort of rounded boss into the cavity, which, in consequence, becomes cup-shaped. As the process enlarges, the upper part of the cup-shaped cavity extends between the rounded central boss and the outer wall, under the form of four canals, which run up parallel with the axis of the process, but stop short of its extremity. Their cæcal ends then send out lateral processes, so as to become T shaped, and ultimately the lateral processes unite together, so as to give rise to a circular canal uniting the ends of the four longitudinal canals. Contemporaneously with these changes, the axis of the boss becomes hollowed out by a canal continuous with the original cavity of the process, and like it lined by the endoderm. A separation now takes place be-

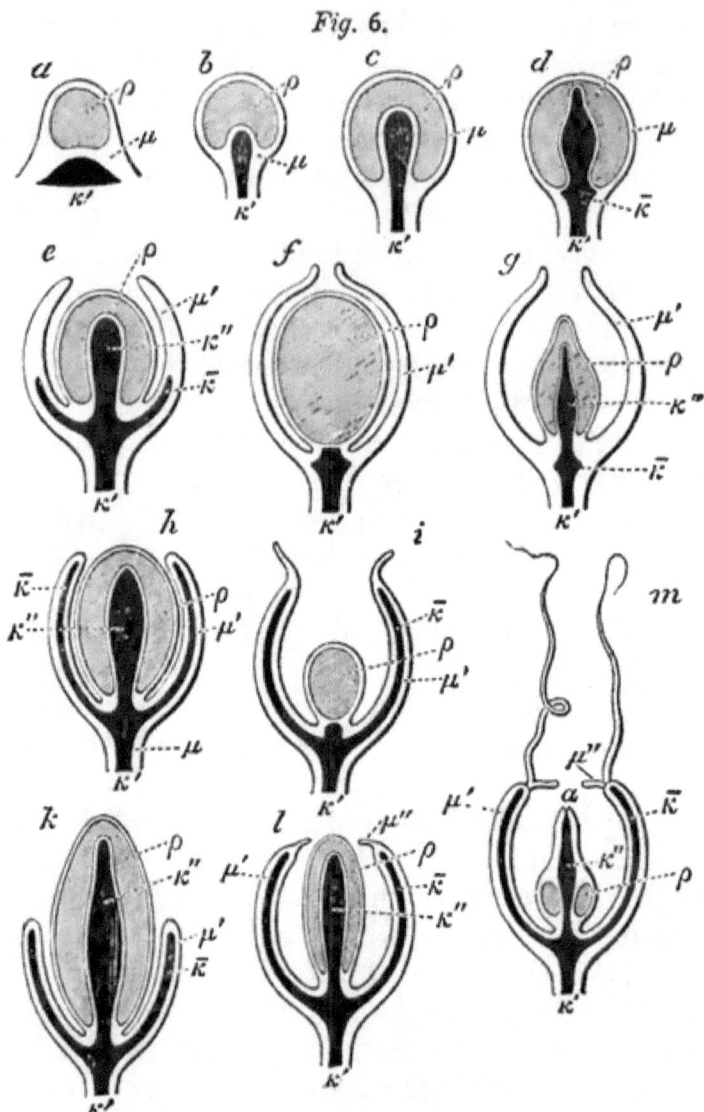

Fig. 6.

Reproductive processes of HYDROZOA:—a—d, simple processes of body-wall, variously modified, containing generative organs; e—m, more complex series of reproductive bodies, each giving rise to a gonocalyx:—μ, portion of body-wall; μ', wall of gonocalyx; μ'', inward projecture of μ', or veil of gonocalyx; κ', simple diverticulum of the somatic cavity; $\bar{\kappa}$, lateral prolongation of the same, forming one of the gonocalycine canals; κ'', cavity of manubrium; ρ, spermaria or ovaria. (These drawings are diagrammatic.)

tween the central and peripheral portions of the thickened boss, commencing at the distal extremity, and extending down very nearly to the proximal end of the boss, so as to leave the thick central portion, enclosing the central cavity, attached to the thin peripheral portion (which remains as the wall of the cavity of the bell) only at its proximal extremity. In the perfect condition of the zoöid thus produced, the endoderm lines the cavity of the peduncle by which it is attached, the canals, and the central cavity of the suspended axial body; while the ectoderm forms the whole of the outer walls of both natatorial organ and central sac."

Still further changes are liable to ensue. The central sac, or 'manubrium,' may acquire a mouth at its distal extremity, thus, as it were, transforming itself into a polypite, while the natatorial organ, or 'gonocalyx,' enlarging, loosens its attachment, and swims freely in the sea as a veritable Medusid (*fig.* 6, *m*). Indeed, there is every reason to believe that a great majority of the organisms described as *Medusidæ* are, in reality, the detached reproductive bodies of other *Hydrozoa* (*figs.* 13 and 14).

Such bodies, however, are more than mere organs. Many of them, when first liberated, present no distinct traces of generative elements, pending the formation of which essential products, their independent existence is of necessity prolonged. At this period they lead a very active life, incre se rapidly in size, and eagerly devour such minute marine animals as they are able to secure. During the calmer seasons of the year they abound in our seas, but before the approach of rough weather usually disappear, their function having been, in all probability, previously discharged. Yet nothing can be more perfect than the series of transitional forms which establish the connection between these highly differentiated organisms and the simple reproductive apparatus occurring in *Hydra*. A few gradations may be indicated. Thus, the closed gonophore of *Cordylophora* sends off from its manubrial cavity a system of prolongations, evidently homologous with true gonocalycine canals (*fig.* 8, *g*). In *Tubularia indivisa*, fully developed canals are exhibited by a distinct gonocalyx, but this never becomes detached (*fig.* 9). Neither does it present marginal tentacles, though even these surround the fixed gonocalyces of *Campanularia Löveni* (*fig.* 10). And so we at once pass to the free swimming generative cups of other *Hydrozoa*. But it would be easy to dwell on further modifications. *Plumularia pinnata*, for example, has its manubrium irregularly lobed, the lobes being, in all probability, as Professor Allman suggests, incipient gonocalycine canals, while in *Campanularia caliculata* canals exist, though the manubrium itself is suppressed. And in some gonophores, the canal system, at first easily recognisable, becomes obliterated by age.

A gonophore, therefore, may exhibit in its development four distinguishable stages, which correspond, respectively, to the permanent forms of the reproductive body in particular members of the group. These conditions are : —

1. That of a simple expansion of the body-wall, as in *Hydra*.

2. That of a well-defined process, or sporosac, as in *Hydractinia*.

3. That of a manubrium with closed gonocalycine investment, in which case the medusoid structure is said to be "disguised," as in the gonophores of *Cordylophora* and numerous other forms.

4. That of a manubrium, with open gonocalyx and well-developed canal system. Such "medusiform gonophores" may either remain attached, as in *Hippopodius* and *Vogtia*, or become free, as in *Velella*, *Porpita*, and many of the fixed *Hydrozoa*.

The same gonophore does not contain more than one kind of generative elements, and these are situated either between the ectodermal and endodermal layers of the manubrium, or in the walls of the gonocalycine canals. When male and female gonophores differ externally in form, the special terms 'androphore' and 'gynophore' are employed to distinguish them. But, apart from such sexual distinctions, two kinds of gonophores appear occasionally to be produced by the same Hydrozoön, while, on the other hand, similar gonophores may arise from the bodies of apparently different species.

So much, then, for the structure of the gonophores; next, as to their position. They may be seated either —

1. on the polypites; or,
2. on special processes termed 'gonoblastidia;' or,
3. directly on the cœnosarc.

The first of these methods is characteristic of

the *Calycophoridæ*, the second of the *Physophoridæ* and *Sertularidæ*, while all three find accessible representatives in the order *Corynidæ*.

In certain species of this last order the gonophores, even on the same individual, obey different modes of attachment. Thus in *Clava multicornis*, some are inserted on the polypites, others on gonoblastidia; while in *Hydractinia*, besides the gonophores borne on the gonoblastidia, a few are found to arise, without intervening support, from the sides of the cœnosarc.

The gonoblastidia are either simple or branched. Often they present a curious resemblance to true polypites, from which, however, they differ in wanting a mouth, and having usually shorter tentacula. Such polypoid gonoblastidia may be examined with ease in *Hydractinia*, where they are less than the polypites in size. In this genus the free extremity of each is seen to end in a pear-shaped, tapering enlargement, whose surface is studded with minute conical swellings containing thread-cells, which increase in size so as to resemble rude tentacles, ten or twelve in number, around the largest portion of the pyriform process. Beneath these the gonophores are borne. At the base of the process is inserted a problematical body, presenting the appearance of a short stalk, which terminates distally in a rounded expansion, filled with very small, dark orange, masses of pigment.

In general, gonoblastidia arise from the sides of the cœnosarc, though, in some cases, they are attached to the bodies of the polypites.

A curious structural modification distinguishes the gonoblastidia of the *Sertularidæ*. In *Cam-*

panularia, for example, columnar gonoblastidia arise in the angles between the stem and branches of the cœnosarc, or from the sides of the branches themselves (*figs.* 10 and 19). The lower portion of each gonoblastidium forms a sort of peduncle, above which the cuticular investment of its ectoderm becomes separated as an urn-shaped capsule, the 'gonotheca.' Such capsules, or "ovigerous vesicles," are very variable and beautiful in form. True gonophores, protected by the gonotheca, are borne along the sides of its axial column.

In some *Calycophoridæ* and *Physophoridæ*, particular regions of the hydrosoma may devote themselves to the performance of the reproductive function, and, becoming separated from the rest of the fabric, subsequently undergo a surprising amount of modification.

Finally, in the *Lucernaridæ*, with the exception of *Lucernaria* and a few closely-allied genera, the reproductive bodies are produced by fission from polypites of almost microscopic minuteness; and, in their detached condition, grow with such rapidity as ultimately to attain a weight of many pounds, or even hundreds. A corresponding advance in structure attends this vast increase of size. Each, at the outset of its free existence, includes a complete transverse segment of the polypite from which it has separated. This soon forms a lobed swimming organ, or umbrella, with the hooded lithocysts before mentioned. From the centre of the umbrella hangs a large polypite, whose lips, in such genera as *Aurelia*, *Cyanea*, and *Chrysaora*, form lobes of considerable length, the folds of which serve as temporary receptacles for the ova during the earlier stages of their development (*fig. 7, b*). The interior of the polypite leads to a large central cavity, situate in the substance of the thick gelatinous umbrella, and

Fig. 7.

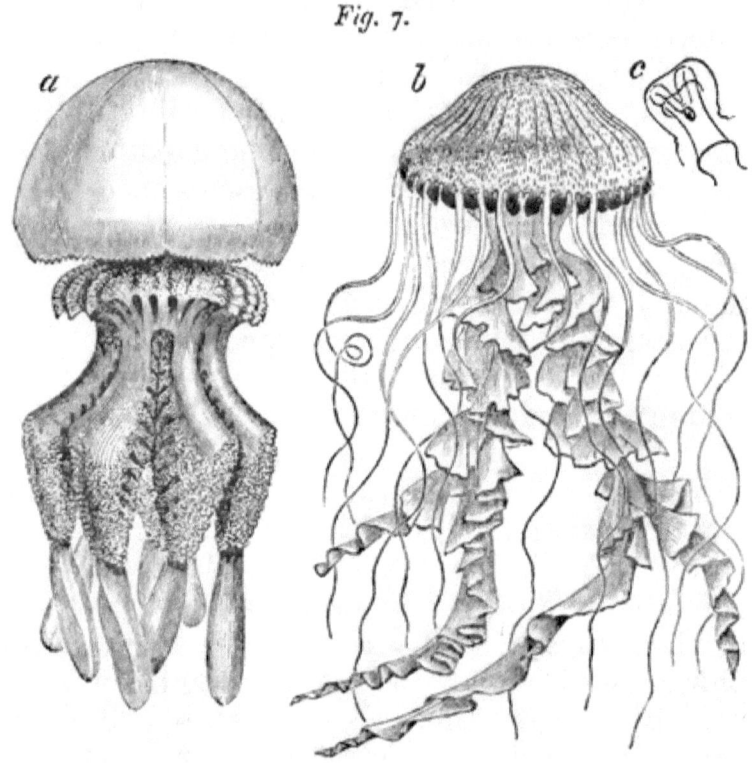

Oceanic forms of LUCERNARIDÆ:—*a, Rhizostoma pulmo; b, Chrysaora hysoscella; c,* its lithocyst. (All reduced.)

lined by a layer of endoderm which sends prolongations into the system of anastomosing canals, communicating with a marginal vessel, fringed, in its turn, by a series of tentacular diverticula. The generative products are lodged in saccular processes of the lower portion of the central cavity, immediately above the bases of the radiating canals, and, being usually of some bright colour, form a conspicuous cross shining through the thickness of the disc.

But in *Rhizostoma, Cephea,* and *Cassiopeia,* a different arrangement prevails, which is best described in the words of Professor Huxley.

"In the *Rhizostomidæ,* a complex, tree-like mass, whose branches, the 'stomatodendra,' end in, and are covered with, minute polypites interspersed with clavate tentacula, is suspended from the middle of the umbrella in a very singular way. The main trunks of the dependent polypiferous tree, in fact, unite above into a thick, flat, quadrate disc, the 'syndendrium,' which is suspended by four stout pillars, the 'dendrostyles,' one springing from each angle, to four corresponding points on the under surface of the umbrella, equidistant from its centre. Under the middle of the umbrella, therefore, there is a chamber whose floor is formed by the quadrate disc, while its roof is constituted by the under wall of the central cavity of the umbrella, and its sides are open. The reproductive elements are developed within radiating, folded diverticula of the roof of this genital cavity" (*fig.* 7, *a*).

This is, without doubt, the most complicated structural product presented by the class, and its description forms a not inappropriate conclusion to the preceding general survey of their organisation.

The majority of *Hydrozoa* are diœcious, the same hydrosoma not bearing both male and female reproductive bodies. Exceptions, however, occur in *Hydra* itself, in *Cordylophora*, in *Plumularia pinnata*, in many *Physophoridæ* and *Calycophoridæ*, *Diphyes* being an exception. The reproductive zoöids of the *Lucernaridæ*, except in the case of *Chrysaora*, appear to be unisexual, but it is not yet ascertained whether generative bodies of dissimilar sexes can be produced by the fission of one primitive hydrosoma.

As in other animals, fecundation is effected by the contact of ova and spermatozoa: the spermaria and ovaria, when fully developed, becoming wholly resolved into these essential elements. The spermatozoa present the form of ovate corpuscles, from the broad end of which a filament projects. The ova' are, in most cases, spherical, destitute of vitelline membrane, with distinct germ-vesicle and germ-spot. Diffusion of the spermatozoa in the surrounding water seems, in the present class, the usual prelude to the act of fertilisation. But, in *Cordylophora*, it has been supposed that the male elements can alone obtain access to the ova by reaching them from within along the general cavity of the body.

SECTION II.

DEVELOPMENT OF HYDROZOA.

The fertilised ovum, in all the *Hydrozoa*, undergoes yolk-division. This process would seem to be determined by the previous division of the germ-vesicle, which, according to Gegenbaur, in some of these animals at least, does not disappear immediately after fecundation.

The embryo which results may be developed from the whole, or only a portion, of the vitellus. It usually appears as a minute, free-swimming ciliated body, but, in some instances, presents a different aspect. The ectoderm and endoderm of the adult Hydrozoön correspond with the inner and outer layers into which the blastoderm of the embryo soon separates, the cavity which is at the same time formed representing the somatic cavity of the future animal.

Hydridæ.—The modification of one end of the body into a hydrorhiza, the formation of a mouth, and of tentacular processes, are the only changes, save those of growth, which seem needed to bring such an embryo into the condition of a perfect *Hydra*. But observations are yet wanting on the development of this organism. The researches of Laurent point to the conclusion that, in the production of the young *Hydra*, a part only of the ovum is directly concerned.

The polypite thus resulting from a true generative act may subsequently, by gemmation, give rise to several others, in all respects similar to the organism from which they were produced. These for a time may remain in connection with each other, but, more usually, they separate, each in its turn, under favourable conditions, to repeat the same budding process. The number of independent beings into which a single *Hydra*, when well supplied with food, and stimulated by a warm temperature, may resolve itself, is certainly astonishing. Not less so are its reparative powers, which seem almost to defy the knife of the anatomist. Full details on this subject are given by Trembley, whose researches on the *Hydra*, published in 1744, are still well worthy of perusal.

Some years ago, Ecker compared the periplastic tissue of the *Hydra* to aggregate masses of the sarcode, or " formless contractile substance," composing the body of *Amœba*. Mr. G. H. Lewes has also recognised distinct "contractile masses," which he says were so very like Amœbæ, as to make him at first believe that the *Hydra* had swallowed them. Such amœboid particles occasionally become detached by the method denominated "diffluence," each usually including one or more endoplasts; but there is good reason to infer that their apparently contractile movements are, for the most part, the result of a process of endosmose. Jäger, however, has shown that two budding Hydræ, each kept by him in a small vessel of water, broke up into several isolated particles, which, after the lapse of a month, were still living, performed amœba-like movements, and, in some instances, passed into a peculiar stage, resembling the encysted condition of *Infusoria*. In this state, Jäger supposes, they may remain throughout the winter, and again, on the return of spring, once more assume the aspect of the primitive *Hydra*.

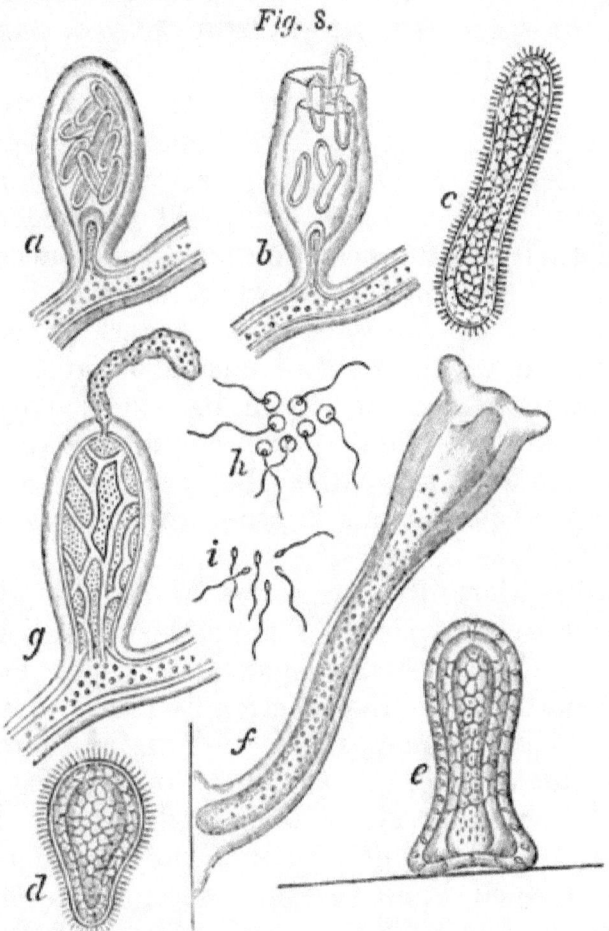

Fig. 8.

Development of CORDYLOPHORA:—*a*, gonophore of *Cordylophora lacustris*, showing embryoes in its interior; *b*, the same more advanced, with embryoes escaping from its ruptured extremity; *c*, an embryo, in its free swimming condition; *d*, the same embryo, having assumed a pyriform figure; *e*, the embryo in its attached condition; *f*, primitive polypite, developed therefrom; *g*, androphore of the same *Cordylophora*, its contents escaping under pressure; *h*, caudate cells liberated therefrom; *i*, spermatozoa. (All magnified.)

CORYNIDÆ. In *Cordylophora*, the free swimming ciliated embryo, on emerging from the ruptured gonophore (*fig.* 8, *b*), is usually of an elongated oval form, but very contractile, so that often it assumes a pyriform figure (*fig.* 8, *c* and *d*). Eventually, the embryo loses its cilia, and, fixing itself, developes a hydrorhiza at one extremity and a mouth at the other, thread-cells being at the same time formed in the ectoderm (*fig.* 8, *e* and *f*). Next, a series of about four tentacles make their appearance; these are soon succeeded by others; the somatic cavity becomes fully formed, and the young *Cordylophora*, increasing in size, is invested with a delicate cuticular layer. The rudimentary cœnosarc, with its single polypite, formed in this manner, soon commences to send forth prolongations, and these, by gemmation, develop the polypites and other appendages of the adult organism.

A somewhat different series of changes occurs in *Tubularia* (*fig.* 9). The embryo of this genus is not ciliated, but first makes its appearance as a discoid body, from the circumference of which short thick processes, the rudiments of tentacula, are produced (*fig.* 9, *f*). The disc then becomes more gibbous at the side turned away from the axis of the gonophore; a mouth, leading into a newly-formed digestive cavity, soon occupying the centre of the opposite side. The mouth then elevates itself on a conical prominence, around which a second series of tentacles arise. In this state the embryo issues from the gonophore (*fig.* 9, *d*). Remaining free for a short time, it finally becomes fixed, and developes a cœnosarc with its cuticular layer (*fig.* 9, *e* and *g*).

Fig. 9.

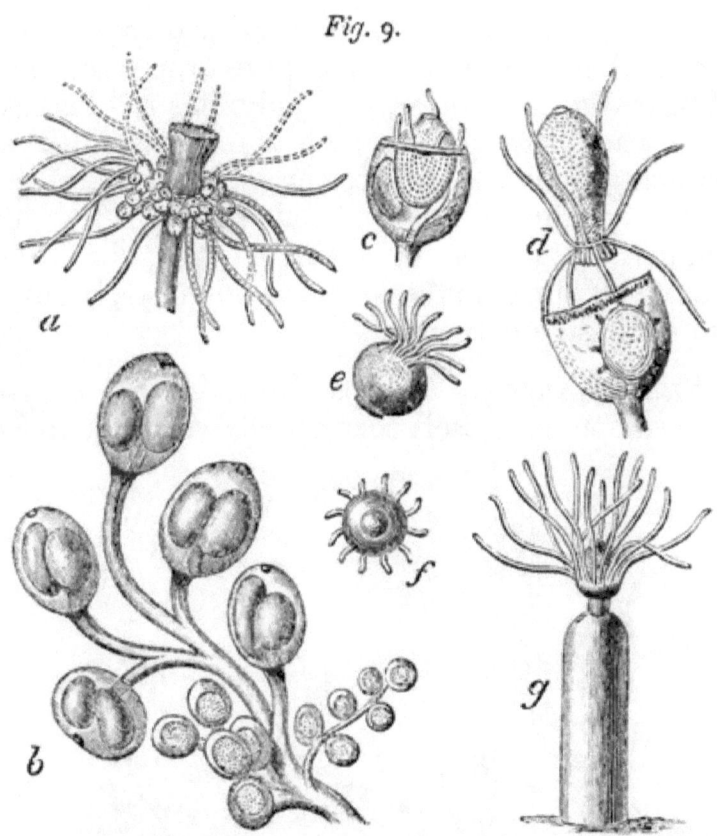

Development of TUBULARIA INDIVISA:—*a*, a polypite, attached to the extremity of the cœnosarc, bearing clusters of reproductive bodies between the two series of tentacles; *b*, one of these clusters consisting of several gonophores, borne on a long branching stalk; *c*, a single gonophore, containing two young polypites, one of which has commenced to extricate itself; *d*, the same gonophore, in a more advanced condition; *e*, a young polypite, thirty-six hours after having escaped from the gonophore; *f*, the younger polypite, shown within the gonophores *c* and *d*; *g*, the polypite *e*, six weeks after extrication. (*a* is about thrice the natural size: the others are much magnified.)

Tubularia, like *Hydra*, and, in all probability, many other *Hydrozoa*, possesses well-marked reparative powers. When living specimens have been kept for some days in vessels of sea-water, it often happens that the polypites drop from their stalks. Soon, however, new polypites are budded forth, each having usually a smaller number of tentacles than its predecessor. Similar are the results produced by artificial fission. In this manner, by section of a single stalk, Sir J. G. Dalyell obtained, in the course of 550 days, twenty-two successive polypites.

SERTULARIDÆ. In most *Corynidæ*, the course of development closely corresponds with that above

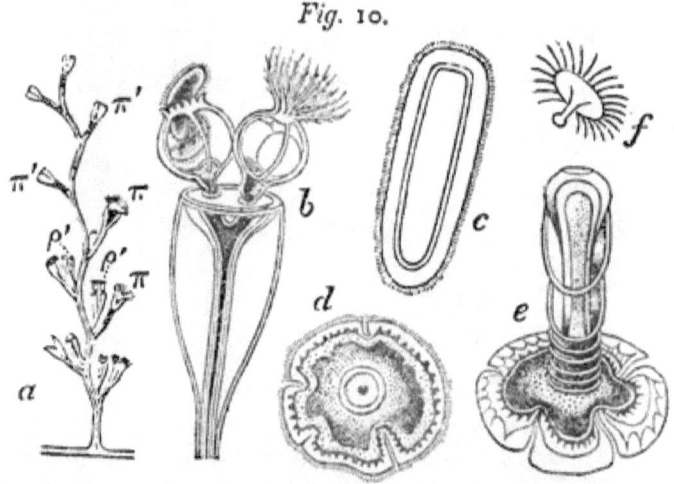

Fig. 10.

Development of CAMPANULARIA:—*a*, *Campanularia Lovéni*; *b*, one of its gonoblastidia from whose summit project two medusiform gonophores, one of which is giving exit to a ciliated embryo; *c*, the same embryo, in its free swimming condition; *d* and *e*, successive stages of the young cœnosarc developed therefrom; *f*, medusiform zoöid of *Campanularia dichotoma*;—π, polypite; π′, hydrotheca; ρ′, gonoblastidium. (*a* and *f* are about twice the natural size; the others are much magnified.)

described as taking place in *Cordylophora*. Of a similar character, in its more general features at least, is the life history of the *Sertularidæ*. The young *Campanularia* or *Antennularia*, at first free, soon loses its cilia, fixes itself, and contracts into a circular disc, which exhibits a division into four lobes (*fig.* 10, *c* and *d*). In the centre of the disc an opaque spot makes its appearance, and over this the surface becomes gradually elevated, until, finally, a young cœnosarc is the result (*fig.* 10, *e*). From this, by gemmation, the branching hydrosoma of the complete organism, with its crowded assemblage of polypites, is subsequently produced.

Thus the young condition of a Sertularid would appear to differ from that of a Corynid in having a portion of its cœnosarc more or less completely developed before distinct traces of a polypite can be observed. Such a conclusion accords well with the composite structure always assumed by the adult hydrosoma. And in this respect the *Sertularidæ*, while departing from the *Corynidæ*, seem to agree with the oceanic orders, *Calycophoridæ* and *Physophoridæ*.

CALYCOPHORIDÆ. Of the earlier embryonic changes in the *Calycophoridæ* little is known. In *Diphyes*, according to Gegenbaur, the blastoderm at first appears as an elevated protuberance, occupying only a portion of the segmented vitellus. Soon, this blastoderm forms a rudimentary nectocalyx, from which a short canal leads to the ciliated cavity of the yolk below. The nectocalyx then rapidly enlarges, while polypites are seen to arise between it and the appended yolk-mass.

PHYSOPHORIDÆ. Our knowledge of the embryology of the *Physophoridæ* is confessedly scanty.

58 HYDROZOA.

The youngest examples of the group seen as yet,

Fig. 11.

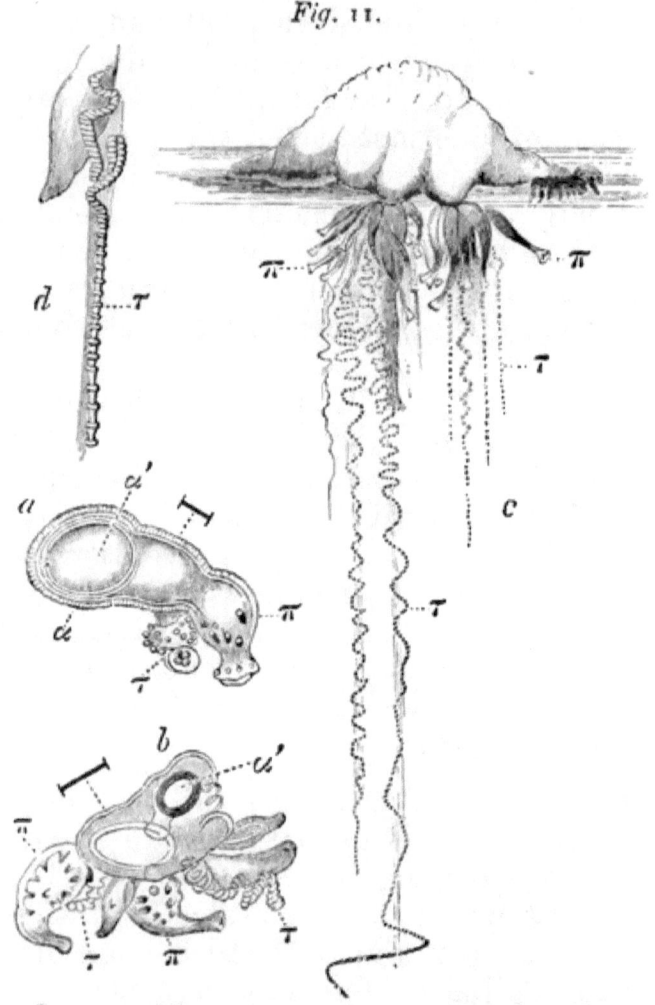

Development of PHYSALIA:—*a*, young *Physalia*, with a single polypite and tentacle; *b*, the same, more advanced; *c*, adult *Physalia*; *d*, a tentacle, with its basal sac;—α, pneumatophore; α′, pneumatocyst; π, polypite; τ, tentacle. (*a* and *b* are magnified; *c* is reduced; *d* is about the natural size.)

exhibit a well-defined pneumatophore, with a single polypite and tentacle (*fig.* 11, *a*).

In a *Velella*, less than ·1 of an inch long, observed by Mr. Huxley, " the horizontal disc of the adult was represented by a bell-shaped, membranous expansion, continued above into a broad crest, half as high as the whole depth of the animal. It was symmetrically disposed, and its superior edge, far from being pointed, was rather concave, and in the centre presented a curious thickening. The central polypite was already open at its distal extremity, and around its base were a few short, cæcal processes, the rudiments of the gonoblastidia or of the tentacles. The margin of the disc was occupied by a single series of large, oval vesicles. The somatic cavity was divided by a series of vertical septa, which passed continuously over the pneumatocyst into the crest, near whose free edge they terminated abruptly, and between them other very short septa were interposed. The somatic cavity and its continuation into the crest were thus broken up into a series of parallel [ciliated] canals, united at their ends by two marginal canals at right angles with one another, one in the disc, the other in the crest. The pneumatocyst shone through the disc, and did not extend into the crest at all." It appeared " as an almost hemispherical body, convex above and flat below. On two of its sides, in a plane perpendicular to that of the crest, there was a double crescentic mark, caused by a depression. The air did not completely distend the pneumatocyst, but appeared to be divided into seven or eight lobes below, so that, at first sight, the organ itself appeared to be lobed, but this was not really the case. It was, in fact, in the smallest specimens a simple vesicle, about ·05 of an inch in diameter, with strong and thin walls, which, when it was burst and the air expelled, fell into sharp folds."

MEDUSIDÆ.—The development of the true *Medusidæ* has yet to be effectively studied. From the observations of J. Müller on *Æginopsis*, of Gegenbaur on *Trachynema* and *Cunina*, and of Fritz Müller on *Liriope*, it seems highly probable that these genera proceed at once from the condition of the embryo to assume the aspect of the organism which gave them birth. Still more conclusive on this point are the results of some recent researches of Claparède on a Medusid closely allied (if not belonging) to the genus *Lizzia*. Within the substance of the body-wall of the de-

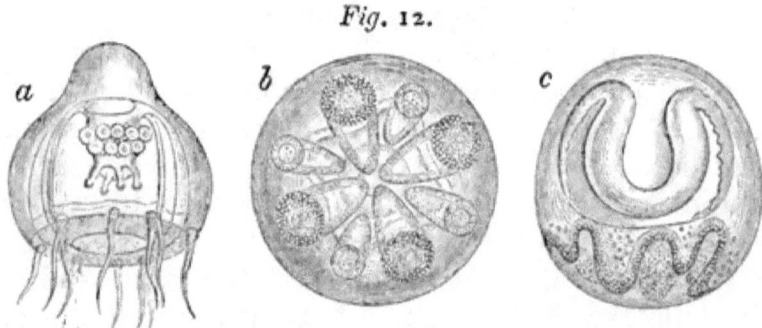

Fig. 12.

Development of LIZZIA:—*a*, adult *Lizzia*, the walls of whose polypite are seen to bear numerous ova; *b*, supposed free-swimming young of *a*, viewed from below; *c*, the same, seen in profile. (*a* is slightly, *b* and *c* are very much, magnified.)

pendent polypite were observed numbers of what seemed to be true ova, some furnished with germ-vesicle and germ-spot, others in a more advanced stage of development. These last resembled in form and structure certain free floating bodies (*fig.* 12), each of which, though still enclosed in an outer covering, presented, on a reduced scale, distinct indications of various parts observed in the adult Medusid; four radiating canals, and eight tentacular enlargements, being especially noticeable. The full-grown *Lizzia* possessed twelve tentacles, four single, and eight others arranged in four alternating pairs. In the young form four of the rudimentary tentacles were much longer than the others, and it seems not improbable that each of them represented one of the four pairs of tentacles in the perfect *Lizzia*. No males of this species have been observed. It is worth adding that another form placed by zoölogists in the same genus appears to be only the detached bud of one of the *Corynidæ*. Yet, as Professor Huxley has said, "it is within the limits of logical possibility that the adult forms anatomically similar, should be genetically different; that they should have arrived at a similar point by different roads."

The observations of M'Crady on the direct development of another Medusid, *Cunina octonaria*, also deserve attention. This creature occurs as a parasite within the nectosac of a distantly allied form, *Turritopsis nutricula*, from whose mouth, by means of a long proboscidiform polypite, the young *Cunina* obtains its food. At an early stage the *Cunina* appears as a clavate body, presenting a short, rudely globular proximal, and a more attenuate, somewhat cylindrical, distal, region. From the sides of the posterior margin of the former, two long flexible tentacles soon sprout, while, at the same time, a nutrient cavity is formed throughout the central portion of the mass. Even at this period the larva produces free buds from its proximal extremity, not more than two appearing to arise at the same time, though the process of gemmation may frequently be repeated. Next, the distal region elongates; the nutrient cavity opens at its free extremity, forming a mouth; and thus a young polypite is produced, while from the proximal margin two new tentacula soon make their appearance. From this region a rudimentary nectocalyx now arises, a fold, in which are developed marginal bodies, appearing, distally, in front of the tentacles, between which four other tubercular lobes are now seen to bud forth. The growth of the nectocalyx slowly proceeds; eight marginal bodies distinctly come into view; the polypite diminishes in size, finally becoming inconspicuous; and the animal attains the adult form characteristic of its family, save only that reproductive organs have not yet been observed.

That instances of the above kind should be multiplied and re-observed seems, for many reasons, very desirable, since, as already remarked, not a few of the forms known as *Medusidæ* are but the free-swimming gonophores of various other *Hydrozoa*. Thus, from the ovum of *Turris*, one of the so-called genera referred to this order, a polypite is produced, which sends forth a creeping cœnosarc, giving rise to a hydrosoma, clearly seem to belong to the *Corynidæ* (*fig.* 13). Dr. T. Strethill Wright has further proved that *Bougainvillea Britannica*, a common form of Medusoid, is, in truth, the reproductive body of *Atractylis ramosa*, one of the *Corynidæ*.

HYDROZOA.

Fig. 13.

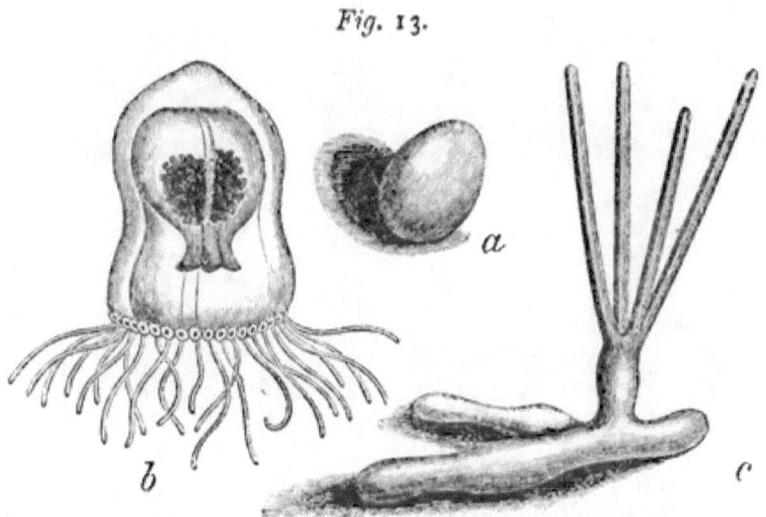

Development of TURRIS:—*a*, ovum of the medusiform zoöid (*b*) known as *Turris neglecta*; *c*, polypite, with creeping hydrorhiza, developed therefrom. (All magnified.)

Certain medusiform gonophores, and, it may be also, some true Medusids, possess the power of producing, by gemmation, free-swimming forms which directly resemble themselves. Such buds have been observed to start from the sides of the polypite in *Sarsia gemmifera* and *Lizzia octopunctata*, from the reproductive region of the calycine canals in two species of *Thaumantias*, from the bases of the tentacles in *Steenstrupia Owenii* and *Sarsia prolifera*, and from the dependent portion of the tentacles themselves in *Diplonema* (*fig.* 14). Hence these medusoids ought, perhaps, to be regarded as free gonoblastidia. Here, also, it may be added, that multiplication by fission has been observed by Kölliker in a species of *Stomobrachium*.

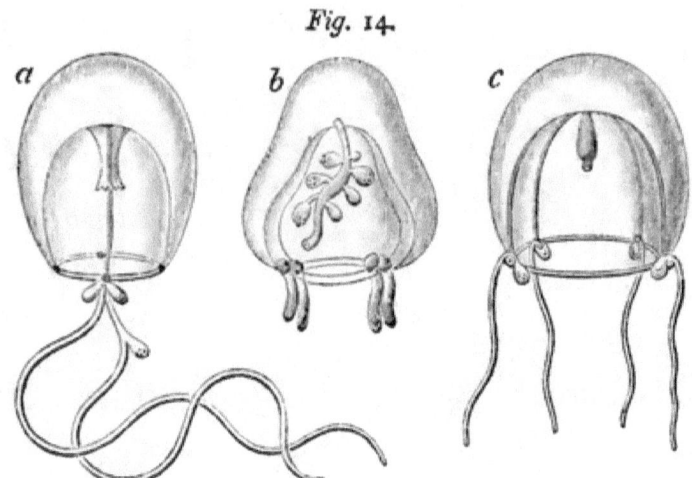

Fig. 14.

Gemmation of MEDUSOIDS:—*a, Diplonema Islandica*, showing young Medusoids budding from the tentacles; *b, Sarsia gemmifera*, with Medusoids arising from the sides of the polypite; *c, Sarsia prolifera*, in which Medusoids are seen to sprout from the junction of the tentacles with the marginal canal. (All magnified.)

LUCERNARIDÆ. Still more singular phenomena appear in the life-history of *Lucernaridæ*. In *Aurelia, Cyanea*, and *Chrysaora*, the ova originate within the generative cavities of the gigantic reproductive bodies previously described. Thence they are transferred, in some unknown manner, to the peculiar pouches formed along the margins of the dependent lips of the polypite, and on their way to these pouches are, in all probability, fertilised by contact with the diffused spermatozoa. Segmentation of the vitellus, and other primordial changes, are undergone by the young ovum while yet within the pouch, from which, about the close of the third day, it comes forth, to enjoy, for a brief period, an active, free-swimming existence. At first it appears as an oblong, flattened, ciliated body, or 'planula,' of very minute size, composed of outer and inner layers, enclosing a central cavity (*fig.* 15, *c*). Soon it assumes a somewhat pyriform figure, enlarging at one extremity, in the centre of which a depression is observable. Next, the narrower end attaches itself to some submarine object, while the depression at the opposite extremity, becoming deeper and deeper, at length communicates with the interior cavity. Thus a mouth is formed, around which may be seen four small protuberances, the rudiments of tentacula (*d*). In the interspaces of these four new tentacles arise; others, in quick succession, make their appearance, until a circlet of numerous filiform appendages, containing thread-cells, surrounds the distal margin of the "Hydra tuba," as the young organism, at this stage of its career, has been termed by Sir J. G. Dalyell (*e* and *f*). The mouth, in the meantime, from being a mere quadrilateral orifice, grows and lengthens itself so as to constitute a true polypite, occupying the axis of the inverted umbrella, or disc, which supports the marginal tentacles. A continuous, wide, open space occupies the whole interior of the umbrella and polypite, whose relations to the rest of the organism, and, indeed, the whole structure of Hydra-tuba, closely resemble what may be seen in *Lucernaria*. Externally, it presents a delicate, translucent aspect, and in height averages some ·3 of an inch. But though dissimilar to *Hydra* in organisation and want of locomotive capacity, the Hydra-tuba recalls to mind its fresh-water congener, first, in its remarkable reparative powers; and, secondly, in the extent to which it multiplies by gemmation. Not merely do buds arise from the sides of the body, but, in addition, creeping tubes, or stolons, are sent forth, from which fresh gemmæ spring up, it may be, to detach themselves, and so one or several large colonies become formed, all the produce of a single fertilised ovum.

For years the hydrosoma may continue in this stage, undergoing no further development. But under certain conditions, similar, perhaps, to those which determine the formation of reproductive organs in the *Hydra*, a new and striking series of changes is inaugurated. First, each Hydratuba elongates, increasing somewhat in size. Then, from just below the tentacles to within a short distance of the proximal extremity, a succession of transverse markings begin to appear, which quickly take on the aspect of circular constrictions (g). When the organism was first discovered in this condition by Sars, he, thinking it a new animal, called it "Scyphistoma." The same naturalist, observing the Scyphistoma at a still later stage, with the constrictions more strongly marked, and the several segments included between them cleft and lobed around their margins, gave it, from its resemblance to an artichoke, the name of Strobila (h). Still further do the constrictions deepen until the Strobila becomes not unlike a pile of cups or saucers. The marginal tentacles then disappear, but a new row arises in their stead from the summit of the short, undivided, proximal extremity (i). The disc-like segments above the tentacles gradually fall off, and, swimming freely by the contractions of the lobed margin which each presents, have been described by Eschscholtz as true *Medusidæ*, under the generic title of Ephyra (k). But each Ephyra soon acquires a nutritive system, lithocysts, tentacles, and generative organs; thus eventually becoming similar to the huge reproductive body, from whose fertilised ovum the primitive Hydra-tuba was produced. This, and the stock which it developed, does not, however, perish, but may again, by growth and fission, give rise to fresh successions of generative bodies.

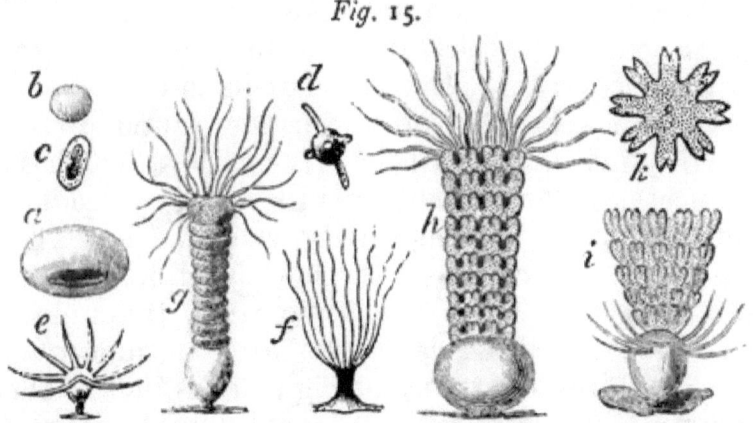

Fig. 15.

Development of CHRYSAORA:— *a*, ova, with gelatinous investment, from *Chrysaora hysoscella*; *b* and *c*, free ova; *d*, young *Hydra-tuba*, with four marginal tentacles, developed therefrom; *e*, the same, with eight tentacles; *f*, *Hydra tuba*, in its ordinary condition; *g*, another *Hydra-tuba*, marked with constrictions; *h*, a more advanced form, with deeper constrictions; *i*, a specimen undergoing fission, in which the tentacles are seen to arise from below the constricted portion, while its upper segments separate, and become free-swimming zoöids (*k*).

Similar to the above appears the life-history of *Cephea* and *Cassiopeia*, notwithstanding the very different structure of the detached reproductive zoöids which these genera present. On the development of *Rhizostoma* itself accurate observations are wanting.

In the *Lucernariadæ* proper, no free zoöids are produced, but the generative elements are formed in longitudinal folds, which arise, opposite to each other, along the four inner angles of the polypite's digestive cavity.

In *Pelagia*, a permanently free form of the same order, the ova are developed directly into the likeness of the organism within which they are evolved. The young *Pelagia*, according to Krohn, presents a minute, semi-transparent, somewhat cylindrical body, invested in a thin, whitish, ciliated, covering. By means of its cilia, the embryo swims rapidly, often turning round on its longitudinal axis. At one extremity, which is truncate, a very small mouth appears, leading into a distinct nutrient cavity, or stomach. This cavity quickly enlarges; the mouth, also, becomes protruded, whilst at the same time, the hinder end of the body is developed into an umbrella. On the third day, traces of eight lobes indent the margin of the umbrella, an equal number of sacs arising from the sides of the stomach. The marginal lobes lengthen, each becomes further indented, and soon rudimentary lithocysts can be distinguished. By this time the oral region is changed into a perceptible polypite, but the organism still moves chiefly by the aid of its cilia, the contractions of the umbrella being at first only occasionally repeated. Afterwards, the ciliated coat disappears; thread-cells are produced; the lips of the polypite enlarge; the umbrella shortens and assumes its proper function; the crystalline contents of the lithocysts make their appearance. The stomachal sacs increase in number and take on the aspect which they present in the adult animal. Finally, marginal tentacles are acquired, four of these being equal in length to the diameter of the swimming organ, while the four other tentacles, and the lips of the polypite, are as yet slightly developed.

The development of the various appendages which arise along the cœnosarc of the composite *Hydrozoa* now requires to be noticed. The same hydrosoma often exhibits these in different stages of evolution, so that their formation admits of being studied, with more or less hope of success, in specimens which may seem to have reached the adult condition.

All the lateral appendages, except the hydrothecæ, appear first as simple processes of the two layers of the body, and in outward form are wonderfully similar, the characteristic aspect of each manifesting itself as growth advances. In the hydrophyllia and nectocalyces the ectoderm enlarges to a much greater extent than the endoderm; in most other appendages, the relations of these layers are not so disproportionate.

As above remarked, there is but slight difference between hydrocysts and polypites at certain stages of their development. But while the hydrocysts remain closed, an opening is formed at the distal extremity of each polypite, villi and distinct regions soon becoming visible in its endodermal lining.

The tentacles, as already shown, differ as to the degree of vacuolation undergone by their endoderm. The lateral threads of branched tentacles " appear successively as close-set buds on one side of the proximal end of the tentacle, the younger buds being always developed on the proximal side of the older ones." When the structure of the branch is complex, consisting of two or three distinct portions, these are gradually produced as each of the bud-like processes elongates. "The involucrum is formed as a process of the ectoderm of the distal end of the peduncle. In *Physophora*, the distal end of the peduncle itself undergoes a singular dilatation, and helps to form the envelope for the sacculus."

The development of the gonophores has, in the account given of their structure, been sufficiently described. That of the nectocalyces is, at first, precisely similar, but the central mass does not, in these, give rise to a manubrium, while the central cavity, into which the longitudinal canals open, remains very much larger.

The hydrothecæ of the *Sertularidæ* are formed by the gradual separation from the body of each polypite of the outer layer, or polypary, excreted by its ectoderm, which, opening distally, displays the cup-shaped cavity, characteristic of different species.

The relative succession of the appendages also demands attention. In the fixed *Hydrozoa* the distal polypites are first developed, whereas the proximal appendages are the youngest in the *Physophoridæ* and *Calycophoridæ*. But this rule does not appear to govern the nectocalyces in the last-mentioned group, their precise order of development still remaining involved in some complexity.

The phenomena indicated in the preceding brief detail of the life-history of the *Hydrozoa* have, in addition to their special value, a wider interest for the philosophic student of zoölogy, from their bearing on the subject of animal development in general. A few words of explanation may therefore, in this place, not appear unnecessary.

The life of every animal species may, from a certain point of view, be regarded as consisting in the alternate performance of two distinct series of acts; the one of reproduction, the other of development.

Each act of reproduction consists essentially in this, that two dissimilar bodies, an ovum and a spermatozoön, are brought into mutual contact. In some cases the spermatozoön penetrates the coats of the ovum, or even enters it by a proper aperture, known as the 'micropyle.'

Thus defined, the process of reproduction is the same in all animals, though in some its simplicity is masked by the occurrence of a variety of other phenomena, all, however, of secondary importance.

It must also be borne in mind that the evolution of ova and spermatozoa, obviously necessary as a prelude to the reproductive function, cannot be considered as forming a part of it. An ovum or spermatozoön is, in truth, nothing more than a highly differentiated portion of the parent organism, the result of a process of development.

But no sooner has the act of reproduction been duly effected, than that of development forthwith begins. The fertilised ovum gives rise to an embryo, which tends to evolve itself into the likeness of its parent. This embryo, together with all the structures subsequently developed therefrom, is said to constitute, in the zoological sense of the term, an animal individual.

Should the resulting organism develop an

ovum in its turn, then, that ovum, if fertilised, forms the basis of a new individual; and so on for every additional ovum concerned in a generative act. So that each performance of the reproductive process is, as it were, the natural boundary between two successive individuals, or, in other words, between two distinct cycles of development. Thus, while the individual perishes, the species, by reproduction, is constantly renewed. Much also, might be said on the analogy which exists between the different individuals of the same species, on the one hand, and the constituent parts of each individual, on the other.

Again, fission and gemmation are not, as many writers incorrectly state, modifications of the reproductive process, but rather, acts of development. For, as already shown, every ovum is at first a bud, which at length, by fission, becomes separated from the body of the parent. All this takes place quite independently of its fecundation. So that an unfertilised ovum is no more entitled to be considered an individual, than a wart or any other excrescence.

The antagonism between development and reproduction, or even between development in general and those particular stages of the vital process by which reproduction is preceded, is sometimes shown by the fact that certain external conditions which seem to favour the one exert an opposite influence on the other. Thus, on the approach of cold weather, the *Hydra* is prone to develop organs of true reproduction, while, if kept in a warm room, it still, as in summer time, continues the formation of ordinary buds.

It is, therefore, the object of the reproductive

function to confer individuality upon that which previously was but a detached part of the parent organism. Howsoever complex the body of an adult animal may seem, it was once an ovum, whose extreme simplicity of structure might almost be said to verge upon homogeneity. What inaugurated the wonderful series of changes by which the ovum fashioned itself into the likeness of its parent? Contact with spermatozoa, or, in one word, reproduction. To say, then, that spermatozoa possess a peculiar individualising influence can scarcely be viewed as a metaphorical form of expression. *How* they are capable of exerting this influence is, however, a problem to which, as yet, science has furnished no definite solution. Bischoff has compared their action to that of a ferment, such as the yeast of beer; but this hypothesis, as Claparède truly observes, only removes the present difficulty a single step backwards.

The zoological individual being, therefore, defined as the entire product of the developmental changes of a single fertilised ovum, we have now to consider the principal modifications which the cycle of development presents.

If all the parts of an individual remain mutually connected, its development is said to be 'continuous'; if any of them separate as independent beings, it is 'discontinuous'.

Continuous development may manifest itself under the three principal modes of 'growth,' 'metamorphosis,' and 'gemmation without fission.' In metamorphosis, growth alternates with certain well-marked changes of form. In gemmation without fission, a tendency to vegetative repetition is more or less distinctly marked. An example of the first of these methods is presented by *Pelagia*; of the second, by *Æginopsis*; of the third, by *Cordylophora* or *Sertularia*.

In discontinuous development the detached portions of the individual are termed 'zoöids,' that which is first formed being distinguished as the 'producing,' that which separates from it, the 'produced' zoöid. If there be more than two successive series of zoöids, the terms 'protozoöid,' 'deuterozoöid,' and 'tritozoöid,' may then be respectively applied to them. Thus, the medusoids budded by *Sarsia* are, probably, tritozoöids. The term zoöid is also extended to the several parts of a connected structure which increases by vegetative repetition; for example, to the polypites, and other appendages of the composite *Hydrozoa*.

The producing zoöid may either possess or want generative organs. In the latter case the produced zoöid may take on the performance of the reproductive function, as in so many orders of *Hydrozoa*.

In this class we have seen that the produced zoöid may resemble the producing zoöid, as in *Hydra*, or be dissimilar to it, as shown by the free-swimming gonophores of the *Corynidæ* and *Sertularidæ*. The first case affords an illustration of simple 'gemmation with fission'; the latter, of the process known as 'metagenesis.' If the producing zoöid possess sexual organs, and the produced zoöid present the morphological, but not the physiological, characters of an ovum, then the process is one of 'parthenogenesis.' All these varieties of discontinuous development are collectively denominated 'agamogenesis,' as distinguished from 'gamogenesis,' in which the ovum, to be developed, must first be brought into contact with spermatozoa.

But such modifications are, in nature, less distinct from one another than the systematic definitions just given might appear to imply. Furthermore, recent investigations on the development of Insects and Crustaceans have tended alike to confuse our old-established notions of animal individuality and of the true nature of the generative process. For certain Insect ova have been observed to undergo development in the ordinary manner, though no previous contact with spermatozoa had taken place. And in unimpregnated female *Vertebrata* ovarian tumours are said sometimes to occur, which contain traces of hair, teeth, bone, nerves, and other tissues proper to the adult organism. If, therefore, cases exist in which the influence of a male element seems rather accessory than essential to the normal evolution of the germ; nay, can even be dispensed with, there are others in which, without such influence, no proper individuality is manifested, though development, to a certain extent, must assuredly be considered to have taken place. For the present we have preferred to advocate the views entertained on these disputed points by Professors Huxley and Carpenter, while, to avoid needless ambiguity, we have thought it better to employ the precise terminology which the former naturalist has suggested.

But other attempts have been made to explain the phenomena in question. Steenstrup, followed in Britain by the late Professor E. Forbes, and a host of minor investigators, proposed to consider both the free zoöids of the *Hydrozoa* and the organisms from which they sprung as alike entitled to the rank of individual beings, belonging to allied groups, and mutually reproducing each other by a process of "alternate generation." But, in addition to the more general objection which may be raised against this hypothesis, confounding, as it does, true generation with gemmation or fission, the one, an act of reproduction, the other, of development, it is sufficient to show that, in the present instance, its application is based on a very superficial examination of the facts to be explained. The gradual series of transitional homologous forms, so surely connecting the complex free gonophores of certain *Hydrozoa* with the simple reproductive processes of *Hydra* or *Hydractinia*, could not have been very familiar to the minds of those who would have hesitated, if called upon, in accordance with Steenstrup's theory, to impute individuality to the latter. Professor R. Leuckart, however, consistently does this, and would regard as true individuals the independent polymorphic buds of the same composite Hydrozoön. And Mr. Lubbock has justly remarked that, "whether we retain the old nomenclature, or dropping the idea of unity in the term '*individual*,' adopt the system proposed by Professor Huxley, we shall be met by great difficulties and inconsistencies." It behoves us, therefore, to follow that explanation which embodies in the simplest manner all the observed phenomena, and which is, at the same time, *least* characterised by inconsistency.

Recently, Professor Agassiz has proposed a modification of Leuckart's theory, and suggests the distinction of four kinds of individuality in the animal kingdom. First, *hereditary* individuality, when from a single egg a single independent being is produced. Secondly, *derivative* or *consecutive* individuality, or "that kind of independence resulting from an individualisation of parts of the product of a single egg;" as in many *Lucernaridæ*, *Corynidæ*, and *Campanulariadæ*. Thirdly, *secondary* individuality, where the product of one egg multiplies by continuous gemmation, giving rise to an immoveable community; as in the *Sertulariadæ*. Lastly, there is *complex* individuality, where a similar but moveable community is formed; as seen in the *Calycophoridæ* and *Physophoridæ*. In this case, he adds, "the individuals of the community are not only connected together, but, under given circumstances, they act together as if they were one individual, while at the same time each individual may perform acts of its own."

Others were for regarding the gonophores of the fixed *Hydrozoa* as the perfect or adult stages of the forms by which they were produced, the whole process being viewed as one of ordinary metamorphosis. The particular objection just stated applies also to the opinion under consideration, which has, nevertheless, found its advocates in a few writers of distinction. There is, no doubt, some degree of plausibility in a view which considers the fixed Corynid or Campanularid as the young condition of the more complex Medusoid to which, by gemmation, it gives rise. It is now, however, certain not only that the *Calycophoridæ* and *Physophoridæ* agree closely in structure with the *Hydrozoa* just named, but likewise, that they bear the same morphological relation to their reproductive bodies. Extend the case to these; let *Velella*, for example, be henceforth the larva of its free medusiform gonophores, and the doctrine which we have contested is at once seen to become untenable.

Another explanation emanated from Professor Owen. His ingenious theory of "parthenogenesis" supposed that the primitive result of each generative act retains within its body unchanged a certain portion of the germ-mass from which it was first evolved, together "with so much of the spermatic force inherited by the retained germ-cells from the parent-cell or germ-vesicle as suffices to set on foot and maintain the same series of formative actions as those which constituted the individual containing them." So that "every successive generation, or series of spontaneous fissions, of the primary impregnated germ-cell, must weaken the spermatic force transmitted to such successive generations of cells." Or, to confine ourselves to the class under consideration, that a Corynid produced, as the resultants of the germ-cells and spermatic force stored up within it, successions of free or fixed gonophores, until the generative force became exhausted. But here, at least, it can be proved, that the unchanged germ-masses alluded to have no objective existence, while the more subjective spermatic force, in these, as in all other, animals, has hitherto succeeded in escaping the ken of the anatomist.

Section III.

CLASSIFICATION OF HYDROZOA.

1. Classification. — 2. Order 1: Hydridæ. — 3. Order 2: Corynidæ. — 4. Order 3: Sertularidæ. — 5. Order 4: Calycophoridæ. — 6. Order 5: Physophoridæ. — 7. Order 6: Medusidæ. — 8. Order 7: Lucernaridæ.

1. **Classification.**—The seven orders into which the class *Hydrozoa* is divided may be defined as follows:

1. *Hydridæ.* — *Hydrozoa*, whose hydrosoma consists of a single locomotive polypite, with tentacles, hydrorhiza, and reproductive organs which appear as simple processes of the body-wall.
2. *Corynidæ.* — *Hydrozoa*, whose hydrosoma is fixed by an hydrorhiza, and consists either of one polypite, or of several connected by a cœnosarc, which usually developes a firm outer layer. Reproductive organs in the form of gonophores, which vary much in structure, and arise from the sides of the polypites, from the cœnosarc, or from gonoblastidia.
3. *Sertularidæ.* — *Hydrozoa*, whose hydrosoma is fixed by an hydrorhiza, and consists of several polypites, protected by hydrothecæ, and connected by a cœnosarc, which is usually branched, and invested with a very firm outer layer. Reproductive organs as gonophores, arising from the cœnosarc, or from gonoblastidia.

4. *Calycophoridæ.* — *Hydrozoa*, whose hydrosoma is free and oceanic, consisting of several polypites connected by a flexible, contractile, unbranched cœnosarc, the proximal extremity of which is furnished with nectocalyces, and dilated to form a somatocyst. Reproductive organs as medusiform gonophores, budded from the peduncles of the polypites.

5. *Physophoridæ.* — *Hydrozoa*, whose hydrosoma is free and oceanic, consisting of several polypites connected by a flexible, contractile, seldom slightly branched, cœnosarc, the proximal extremity of which expands into a pneumatophore, and is sometimes provided with nectocalyces. Reproductive organs, more or less complex in structure, developed upon gonoblastidia.

6. *Medusidæ.* — *Hydrozoa*, whose hydrosoma is free and oceanic, consisting of a single polypite suspended from the roof of a nectocalyx, furnished with a system of canals. Reproductive organs as processes either of the sides of the polypite, or of the nectocalycine canals.

7. *Lucernaridæ.* — *Hydrozoa*, whose hydrosoma has its base developed into an umbrella in the walls of which the reproductive organs are produced.

The characteristics of these orders are indicated more briefly in the subjoined analytical table.

2. **Order 1: Hydridæ.** — The order *Hydridæ* contains but a single genus, *Hydra*, distinguished from the few marine *Hydrozoa* which it approaches in physiognomy by its peculiar habit and locomo-

HYDROZOA.

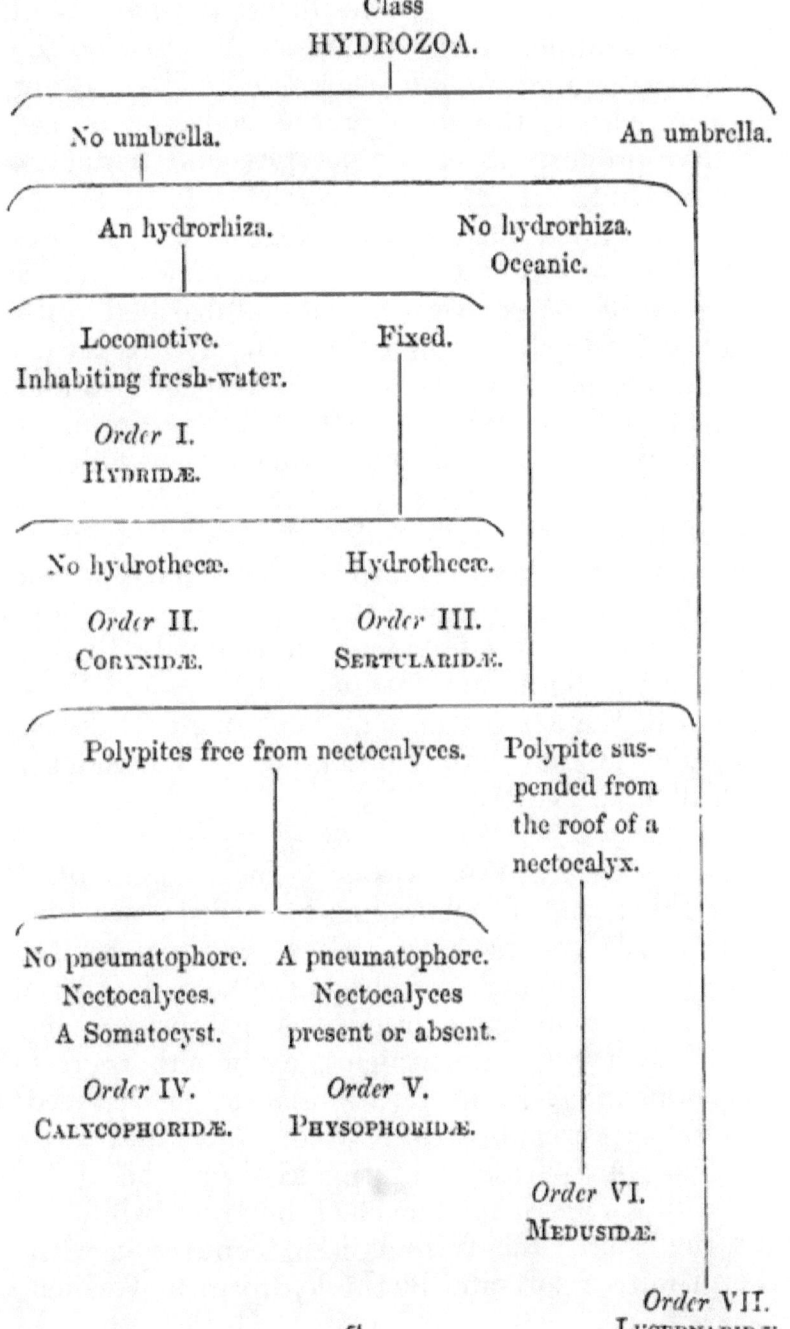

tive powers. Several species of *Hydra* have been described under such names as *H. viridis*, *H. rubra*, *H. vulgaris* and *H. fusca*. These differ in size, colour, the form of the body, or in the relative proportions of the polypite and tentacles. The polypite of *H. vulgaris* is cylindrical, its colour variable, but usually orange-brown, and its tentacles of moderate length. *H. viridis* has a polypite of a grass-green tint, furnished with comparatively short tentacula. *H. fusca* is larger than either of these, its colour is deep brown, and its tentacles very long and extensile; the proximal extremity of the polypite becoming suddenly attenuated for about a third of its length. When living Hydræ are removed from the water, they appear to the eye as minute specks of jelly, which quickly, however, recover their true form on re-immersion. In confinement they readily thrive, seeking the light and feeding voraciously. Specimens of the *Hydra* may be kept in glass vessels, and their singular habits observed by the student, with little difficulty.

3. **Order 2: Corynidæ.** — In *Corymorpha*, *Vorticlava*, and *Myriothela*, the hydrosoma, like that of *Hydra*, presents only a single polypite, but, in the greater number of *Corynidæ*, it is composite, exhibiting numerous polypites connected by a cœnosarc, which may be either erect and branching, as in *Cordylophora*, or reduced to a delicate creeping tube, as in *Clava* and *Trichydra*. A hydrosoma of this kind may be compared to a *Hydra* in the act of budding, while as yet the young zoöids remain in connection with the primitive polypite, by the hydrorhiza of which

the entire fabric continues to attach itself. And, since gemmation may take place in many different ways, so, in like manner, result the great variety of forms due to modifications of what is essentially the same process of growth. In the flower-like *Tubularia indivisa*, the cœnosarc consists of several simple tubes, intertwined one with another near their attached extremities, and sometimes rising to a height of ten or twelve inches (*fig.* 16). From the distal ends of these tubes, which are of a straw-yellow colour, the polypites, tinged with a bright scarlet, conspicuously project. The hydrosoma of *Eudendrium rameum*, though seldom more than six inches in height, bears a singularly close resemblance to a forest-tree in miniature, its surface being studded with minute reddish polypites, not less than a thousand of which may crowd the branches of a single specimen. Such arborescent structures strikingly contrast with the slender mossy threads which compose the connecting stem of smaller species. In *Hydractinia*, the meshes of the very intricate creeping cœnosarc are aggregated so as to form a compact lamina or crust, investing the surfaces of univalve shells, which, by a coincidence hitherto unexplained, usually afford shelter to the Hermit-Crab. The living coating of *Hydractinia* presents to the naked eye the appearance of a rather coarse flocculent nap, pale grey or milk-white in tint, the polypites, when fully expanded, attaining a height of nearly half an inch, and waving to and fro with every agitation of the shell (*fig.* 17, *c.*) In other genera the colour of the cœnosarc is usually yellowish-brown. All the preceding forms have the hydrosoma rooted, or attached to other objects, and in no Corynid hitherto observed does it appear to be altogether free, unless, indeed, an exception

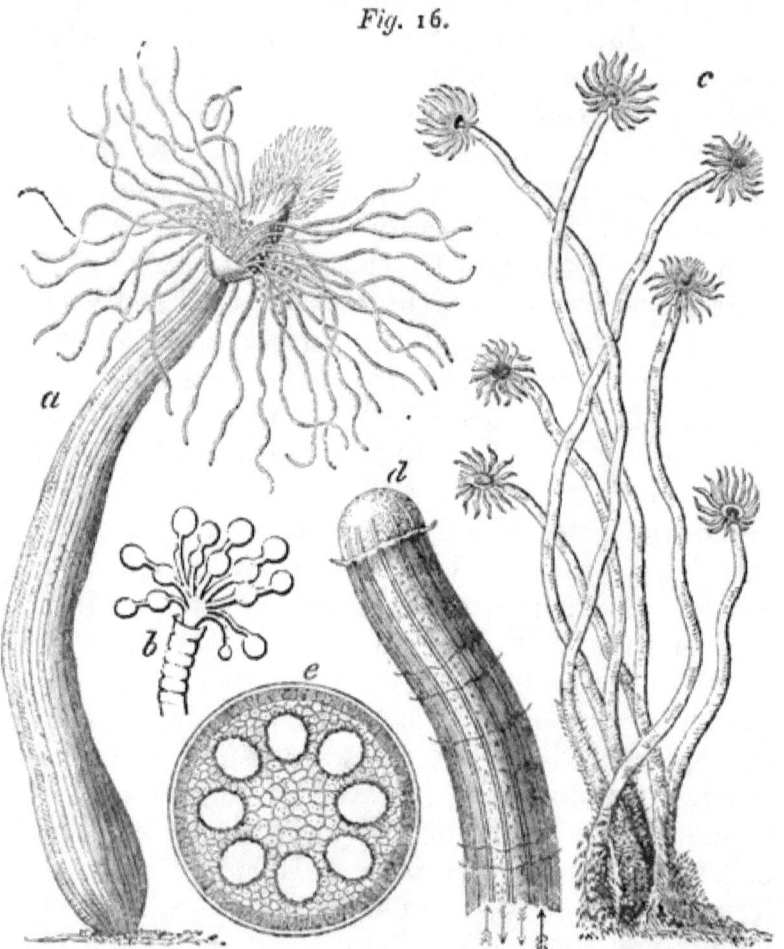

Fig. 16.

Morphology of TUBULARIADÆ:—*a*, *Corymorpha nutans*; *b*, tuft of gonophores from *Corymbogonium capillare*; *c*, *Tubularia indivisa*; *d*, distal extremity of its cœnosarc; *e*, transverse section of the same. (*a* and *c* are of the natural size; *b*, *d*, and *e*, are magnified.)

be made in favour of the doubtful genus *Nemopsis*.

The firm horny layer, or polypary, which the cœnosarc excretes in *Tubularia* and its allies, remains in a comparatively rudimentary condition among most other *Corynidæ*. In few, however, is it absent altogether. In *Hydractinia*, it becomes elevated at intervals to form numerous rough processes or spines, while over the general surface of the ectoderm its presence is almost imperceptible. A very different modification is presented by the genus *Bimeria*. Here the polypary is not, as in other members of the order, restricted to the cœnosarc, but extends itself so as to clothe the entire body of each polypite, leaving bare only the mouth and tips of the tentacles.

The chief differences which prevail among the polypites of the *Corynidæ* have reference either to size or the disposition of their tentacula. The comparatively gigantic polypite of *Corymorpha nutans*, which attains a length of 4·5 inches, is described by Forbes and Goodsir as presenting the appearance of a beautiful flower, nodding gracefully upon its stem (*fig.* 16, *a*). Another species of the same genus, *C. nana*, does not exceed ·5 of an inch in length, though this, in its turn, is double the size of the tiny *Vorticlava humilis* (*fig.* 17, *a*). Still more minute are the delicate polypites of some species of *Eudendrium*. In most members of the present group the general form of the polypite is more or less clavate.

The tentacles exhibit several distinct modes of arrangement. In *Tubularia* and *Corymorpha* a fringe of short appendages immediately surrounds the mouth of the polypite, from the base of which, close to the distal extremity of the cœnosarc, arises a second circlet of much longer filiform tentacula, fewer in number than those of the upper row (*fig. 9, a*). *Vorticlava*, also, possesses a twofold series of tentacles, but here, those of the lower circlet are twice as numerous as the upper, which are five in number, short, stout, and capitate (*fig.* 17, *b*). In *Clava*, *Cordylophora* (*fig.* 5, *c*), and *Coryne* (*fig.* 17,

Fig. 17.

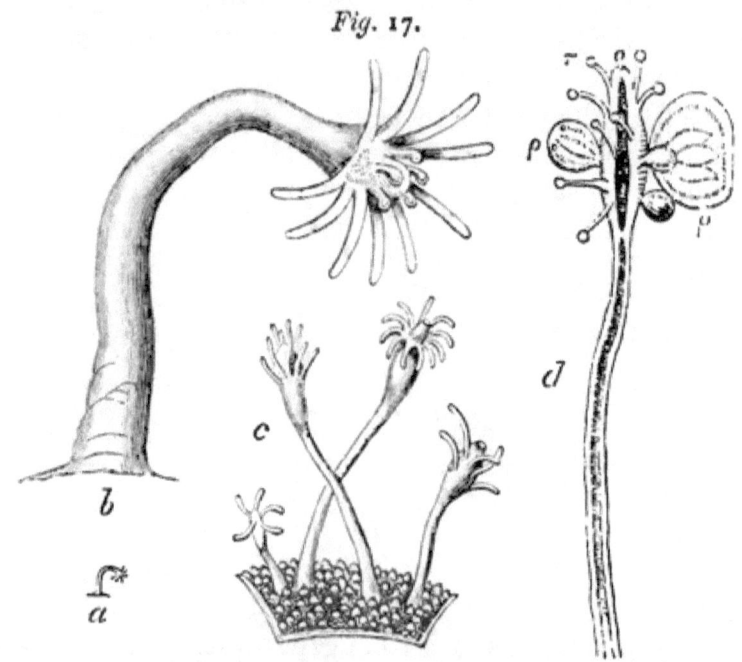

Various forms of CORYNIADÆ:—*a* and *b*, *Vorticlava humilis*; *c*, four polypites of *Hydractinia echinata*, growing on a piece of shell; *d*, portion of *Syncoryne Sarsii*, with medusiform zoöids (*p*) budding from between the tentacles (*τ*) of the polypite (*o*). (All, except *a*, magnified.)

d), the tentacles appear irregularly scattered along the sides of each polypite, though most abundantly towards its distal extremity. In *Coryne* the mouth is highly flexible, possessing the power of bending towards that tentacle which has seized the prey, and of converting itself, upon occasion, into a kind of sucking disc. The polypites of the allied genus *Stauridia* are distinguished by the possession of two or more cycles of dissimilar tentacles, separated from one another by a considerable interval, each cycle including four tentacles; the lower row filiform, the upper whorl, or whorls, capitate, and placed at right angles to one another and the polypite. In *Pennaria*, there is a basal circlet like that of *Tubularia*, between which and the mouth of the polypite lie scattered numbers of shorter tentacles, resembling those of *Vorticlava*. In *Myriothela*, multitudes of wart-like tentacles crowd the whole surface of the club-shaped, solitary, polypite. Similar to these are the tentacles of *Acaulis*, which exhibits, in addition, a basal series of long prehensile appendages. These, however, disappear as the organism approaches maturity, so that this form may possibly be but a young condition of *Myriothela*. A single series of rather long tentacles, inserted as in the freshwater *Hydra*, arises at a short distance below the mouth in *Trichydra*, *Clavatella*, *Perigonimus*, *Bimeria*, and *Eudendrium*. Some species of these genera seem to foreshadow an arrangement of the tentacles which in *Hydractinia* becomes sufficiently conspicuous. Around the mouths of the digestive zoöids in this genus two rows of alternating tentacles are placed, so close to each other that they appear, at first sight, to constitute a single series; the lower tentacula, which are shorter, projecting at right angles to the body of the polypite, from the axis of which the upper tentacles very slightly diverge. These, however, have no direct connection with the long tentacular appendages, arising directly from the cœnosarc, to which allusion has already been made. Lastly, in *Lar*, each polypite supports but two tentacles, above which the mouth is furnished with a pair of wide projecting lobes, capable of being approximated closely to each other, and serving, doubtless, as efficient organs of prehension.

The gonophores of the *Corynidæ* vary not a little both in structure and mode of attachment. In *Cordylophora, Perigonimus, Garveia, Bimeria,* and some forms of *Eudendrium* and *Atractylis,* they spring directly from the stem or branches of the cœnosarc. In other species of the two last-mentioned genera they are seated either beneath the tentacles of the polypites, or on the summits of special branches, arising from the proximal region of the hydrosoma (*fig.* 16, *b*). In *Myriothela, Acaulis,* and *Clavatella,* the gonophores have their origin on the polypite, not far from its attached extremity: in *Coryne* and *Stauridia,* they are produced between the tentacles (*fig.* 17, *d*). In *Corymorpha* and some species of *Tubularia,* they are supported on long branching gonoblastidia, inserted immediately within the basal circlet of tentacula (*fig.* 9, *b*): in other *Tubulariæ, T. calamaris* and *T. Dumortierii,* as also in the genus *Pennaria,* these long stalks appear to be absent. The arrangement of the reproductive bodies in *Clava* and *Hydractinia* has already been pointed out. In the closely allied genera, *Dicoryne* and *Podocoryne,* they originate, in a somewhat similar manner, on proper gonoblastidia, never on the ordinary polypites. But the proliferous stalks of *Podocoryne* are furnished, each, with a mouth, and differ little from true polypites save in their smaller size and the possession of fewer tentacula.

Transitional forms of this kind should not, however, surprise us, when we consider the common bond, community of descent, which connects the two kinds of appendages in question.

In the genera *Perigonimus, Atractylis, Pennaria, Corymorpha, Acaulis, Stauridia*, and some forms of *Coryne, Tubularia*, and *Eudendrium*, the gonophores assume the aspect of free-swimming medusoids. In most other *Corynidæ* they are fixed, exhibiting many remarkable gradations of structure. An intermediate condition is presented by the curious reproductive zoöids of *Clavatella*, which, though locomotive, scarcely merit the appellation of medusiform. They are described by Mr. Hincks as free polypoid buds, furnished with six forked processes, set round the margin of a central hemispherical disc, one limb of each fork being capitate, like the tentacles of the polypite itself, the other terminating in a peculiar sucker-like enlargement. By means of these organs the zoöid, when detached, moves freely about, until finally it proceeds to mature its generative products.

The gonophores of *Trichydra, Vorticlava*, and *Lar* have hitherto remained unknown.

Two families of *Corynidæ* have been distinguished, though the character employed to separate them appears to be somewhat artificial.

Order CORYNIDÆ.

Family 1. CORYNIADÆ.
Polypary absent, or rudimentary.

Family 2. TUBULARIADÆ.
Polypary well developed.

The entire order is sometimes denominated *Tubularidæ*, and agrees with the group *Tubularina* of Ehrenberg.

4. **Order 3: Sertularidæ.** — Like the members of the preceding order, all the *Sertularidæ*, after the expiration of their embryonic condition, become permanently fixed by means of the hydrorhiza which forms the proximal extremity of the cœnosarc (*fig.* 4, *c*). In this group the tendency to increase by gemmation is even greater than among the *Corynidæ*, for no example of a Sertularid has yet been recorded in which the hydrosoma exhibits but a single polypite. The cœnosarc is plant-like and, frequently, much branched, the main stem either losing itself in its own ramifications or remaining distinct throughout the entire length of the arborescent mass. A good example of the latter mode of growth is afforded by the Sea-Fir, *Sertularia cupressina*, the hydrosoma of which may attain a height of two, or even three, feet, and bear on its branches so many as 100,000 distinct polypites. In contrast with this, the largest of our native species, may be mentioned the delicate *Sertularia tenella*, the length of whose slender creeping hydrosoma scarcely reaches one inch. The waving fronds of Oar-weed on various parts of the coast afford a suitable habitat to the anastomosing thread-like cœnosarc of another characteristic species, *Campanularia geniculata*, which sends up at intervals its peculiar zig-zag branches, from the angles of which the polypite stalks arise. Other *Sertularidæ* attach themselves to stones or shells, and not a few of the smaller forms occur parasitically on the stems of more conspic-

uous species. Examples of this habit are afforded

Fig. 18.

Morphology of SERTULARIADÆ:—*a*, dried hydrosoma of *Sertularia tricuspidata*; *b*, portion of the same; *c*, fragment of the cœnosarc from a dead specimen of *Halecium halecinum*; *d*, gonoblastidium of *H. Beanii*; *e*, three gonoblastidia of *Sertularia argentea*;—π', hydrotheca; κ, cœnosarc; ρ', gonoblastidium. (All, except *a*, magnified.)

by the genera *Coppinia* and *Reticularia*, and by several of the true *Sertulariæ*.

The cœnosarc, in all cases, excretes a very firm chitinous polypary, usually of a pale horny colour, which may either remain throughout in close contiguity with the ectoderm, or become separated from it at regular intervals, so as to impart an elegant ringed appearance to portions of the tree-like structure (*fig.* 19, *b*). This semi-transparent, horny, sheath persists long after the destruction of the soft parts of the organism, so that, among the larger species of *Sertularidæ*, the peculiar form of the hydrosoma is sufficiently well seen in dried specimens. Here, therefore, the polypary differs from that of the *Corynidæ* in its firmer texture, but the most important distinctive feature of the present order is found in the occurrence of the hydrothecæ; organs which do not exist in any other group of *Hydrozoa* (*fig.* 18, *b*). The nature of these appendages has already been explained. Their numerous, often beautiful, diversities of form and mode of arrangement afford aids to the definition of the minor types of structure which occur within the limits of this circumscribed group. In *Campanularia fastigiata* the distal end of the hydrotheca forms, according to Mr. Alder, a sort of "operculum, which, when closed, slopes down on each side like the roof of a house, the two opposite angles forming the gables. When the operculum is fully open, the folds disappear, and the edges unite into a continuous rim round the top of the cell."

The polypites of the *Sertularidæ*, more minute than those of the *Corynidæ*, differ little from one another, either in form or the general arrangement of their tentacles. In *Sertulariadæ* proper the polypites are sessile, while in the *Campanulariadæ* each is elevated on a conspicuous stalk. An intermediate condition is presented by the genus *Halecium*, the polypites of which are 'sub-sessile,' each hydrotheca being jointed to a short process of the cœnosarc (*fig.* 18, *c*).

The tentacles, though apparently disposed, Hydra-like, in a single row below the mouth, are found, on close examination, to exhibit an indistinct alternate arrangement; slight differences in length and position distinguishing those of the two series. The peculiar rough appearance which each tentacle presents resolves itself under the microscope, into rows of minute elevations, or 'palpocils,' within which numbers of thread-cells are lodged. The tentacles are filiform, tapering gradually towards their free extremities. In *Campanulina* a delicate web-like extension from the body of the polypite unites these appendages for about a sixth of their entire length.

Allusion has elsewhere been made to the nematophores, or characteristic organs of offence, noticed by Mr. Busk in the genus *Plumularia*, and one or two of its immediate allies. These singular appendages are well deserving of minute investigation. Their offensive nature seems proved by the abundance of thread-cells in their interior, coupled with the fact that certain species of *Plumularia* have been observed to sting with some severity. In *Plumularia* proper one of these organs arises on either side of each hydrotheca, while in *Halicornaria* they are situated, between the polypites, on the general surface of the cœnosarc.

The reproductive organs vary, perhaps, less than those of the *Corynidæ*, and are usually supported on the curiously modified gonoblastidia, whose structure has previously been described. Among the *Campanulariadæ* they frequently assume the form of free-swimming medusoids;

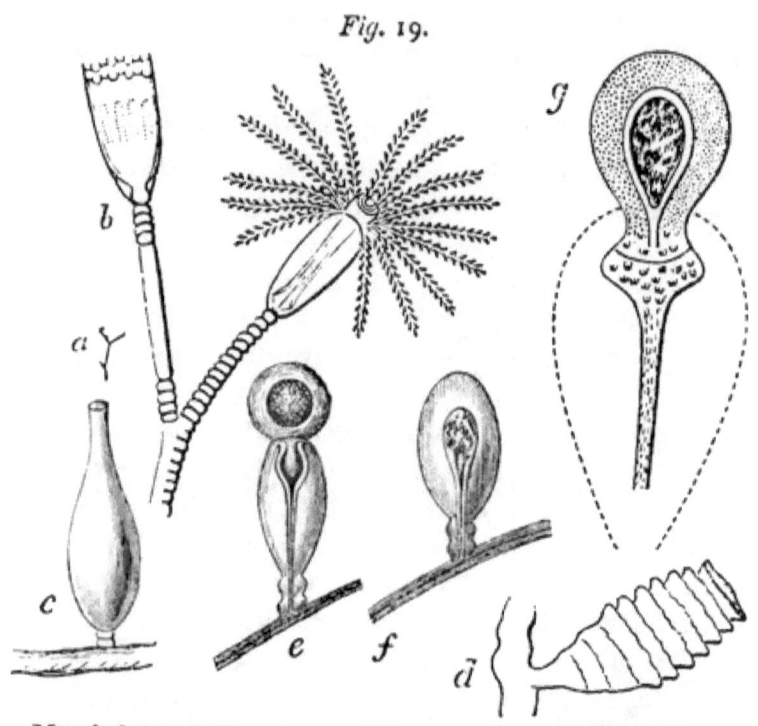

Fig. 19.

Morphology of CAMPANULARIADÆ:—*a*, *Laomedea neglecta*; *b*, portion of the same; *c*, gonoblastidium of *Campanularia volubilis*; *d*, gonoblastidium of *C. Johnstoni*; *e*, gonoblastidium of *C. syringa*; *f*, the same in an earlier stage; *g*, upper portion of *e*, slightly compressed. (All, except *a*, magnified.)

but in the *Sertulariadæ* seldom, if ever, become detached.

In some *Plumulariæ* the gonophores appear to be naked. In *P. cristata* the branch bearing these organs undergoes a curious metamorphosis by the development from its opposite sides of alternate leaflets, which eventually arch over, and unite with one another, forming a basket-like receptacle, or 'corbula,' within which the reproductive bodies are lodged.

In *Sertularia polyzonias* and some other species only one gonophore, consisting of a simple closed sac, arises from the gonoblastidial column, and, by the protrusion of this sac beyond the orifice of the urn, an external capsule, or 'acrocyst,' is formed, into which the ova are transferred at a certain period of their development (*fig.* 19, *e*, *f*, and *g*).

In *Campanularia Löveni* the ripe gonoblastidium displays at its summit the medusa-like gonophores already alluded to, whose form is, in many respects, so peculiar that Professor Allman has proposed to designate them by a distinct name, 'meconidia' (*fig.* 10). The reproductive elements of this species are developed, as in the *Corynidæ*, between the ectoderm and endoderm of the manubrial wall, while in other *Sertularidæ*, with medusa-like gonophores, they arise in the course of the calycine canals.

The order *Sertularidæ* includes two families.

Order SERTULARIDÆ.

Family 1. SERTULARIADÆ.
Hydrothecæ, and polypites, sessile.

Family 2. CAMPANULARIADÆ.
Hydrothecæ, and polypites, stalked.

A more extended acquaintance with the position of the nematophores may perhaps afford grounds for modifying this arrangement.

5. Order 4: Calycophoridæ. — The members of the next order, *Calycophoridæ*, appear, at first sight, very dissimilar in aspect to the fixed *Hydrozoa*, which, nevertheless, in all essential characteristics, they closely resemble. *Diphyes*, the type of the group, presents a delicate filiform cœnosarc, to the proximal extremity of which are attached two large, firm, mitrate, nectocalyces (*fig.* 20, *a*). To these appendages, which differ slightly in form, the distinctive terms of 'proximal' and 'distal' have been assigned. The former, as its name imports, precisely terminal in position, is furnished with a conical cavity running parallel with, but distinct from, its nectosac. Into this cavity is fitted the apex of the distal nectocalyx, along the inner surface of which it prolongs itself as a lengthened groove, with its sides arched over in such a manner as to form a more or less perfectly closed canal. The cœnosarc, with its numerous appendages, freely glides up and down the peculiar chamber, or 'hydrœcium,' thus produced, into which it can, upon occasion, be completely retracted. The cœnosarc itself dilates slightly towards its proximal extremity into a small ciliated chamber, which, narrowing above, becomes continuous with a sac of larger size, termed the 'somatocyst.' This, too, is ciliated, its cavity appearing in most cases almost obliterated through excessive vacuolation of the endoderm. The somatocyst is firmly embedded in that portion of the proximal nectocalyx which forms the upper boundary of the hydrœcium, while from the smaller ciliated chamber two ducts are given off, one to the distal, the other to the proximal nectocalyx, where each communicates with the small cavity common to the nectocalycine canals. Along the sides of the cœnosarc are placed the several appendages, con-

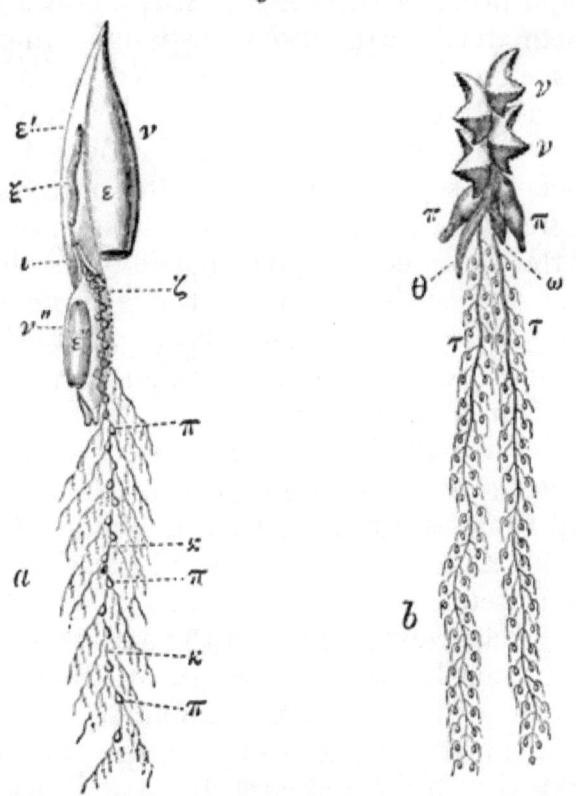

Fig. 20.

Morphology of CALYCOPHORIDÆ:— *a*, *Diphyes appendiculata*; *b*, *Vogtia pentacantha*;—*v*, proximal nectocalyx of *Diphyes*; ϵ', its posterior contour; ϵ, its nectosac; v'', distal nectocalyx; ϵ'', its nectosac; ξ, somatocyst; ι, proximal portion of hydrœcium; ζ, proximal extremity of cœnosarc; κ, its distal extremity; π, polypite, with its tentacle; v, v, nectocalyces of *Vogtia*; π, π, its polypites; τ, τ, their tentacles; θ, androphore; ω, gynophore. (Natural size.)

sisting chiefly of polypites, tentacles, hydrophyllia, and organs of reproduction. Large specimens of

Diphyes attain, when fully extended, a length of several inches, their cœnosarc giving support to at least fifty distinct polypites. Of the great beauty of these, and other oceanic *Hydrozoa*, no description can adequately treat. So transparent, in many cases, is the delicate cœnosarc, that its course upon distant inspection is revealed only by the bright tints of some of its appendages. A touch is often sufficient to separate it from the nectocalyces, which, from their size and firm consistence, constitute the most conspicuous portions of the organism. Hence the origin of the generic name, *Diphyes*, devised by Cuvier, who regarded the two swimming organs as distinct animals, imperfectly united with one another.

An unbranched, filiform, cœnosarc occurs in all *Calycophoridæ*. In *Hippopodius* its proximal extremity folds inwards to form a loop, so that the true position of the nectocalyces is thereby somewhat confused.

Of the many appendages to the cœnosarc by far the most remarkable are those just mentioned. In accordance with the relative number, structure, and arrangement of these organs, the few genera of the order hitherto carefully examined may readily be identified and separated from one another; as shown in the accompanying table.

ARTIFICIAL ARRANGEMENT OF CALYCOPHORIDÆ.

1. { Nectocalyces two in number 2
   { Nectocalyces numerous, similar 4
   { A single, proximal, spheroidal, nectocalyx . *Sphæronectes*.
2. { Nectocalyces unlike in size and form 3
   { Nectocalyces similar . . . . . *Praya*.
3. { Proximal nectocalyx equal to, or larger than, the distal one . . . . . *Diphyes*.
   { Proximal nectocalyx shorter than the distal *Abyla*.

4. {
- Nectocalyces horse-shoe shaped . . . . *Hippopodius.*
- Nectocalyces concave externally, "and produced into five points of which the three upper are much longer and stronger than the two lower." . . . . . *Vogtia.*

Praya, Hippopodius, and *Vogtia* have 'incomplete' hydrœcia, the nectocalycine groove along which the cœnosarc glides not forming, in these genera, a closed canal. In *Praya,* however, the two, nearly symmetrical, terminal nectocalyces have their open grooves so applied to each other as to form, by their apposition, a short tube (*fig.* 4, *d*).

The polypites and tentacles of the several genera of *Calycophoridæ* present no very striking differences of structure.

Not so, however, the hydrophyllia. *Abyla,* the genus most closely allied to *Diphyes,* is distinguished from that form not merely by its nectocalyces, but also in having thick, facetted, hydrophyllia, the edges of which do not overlap one another. In *Diphyes* the hydrophyllia are foliaceous, smooth externally, slightly convex, and folded so that their edges freely overlap.

In *Praya,* "each hydrophyllium is a thick, gelatinous, and reniform body, bent upon itself, rounded and solid at one extremity, and divided at the other into a median thick and two lateral lamellar lobes. The phyllocyst is prolonged into four cæcal processes." But in *Vogtia, Hippopodius,* and, perhaps also, *Sphæronectes,* these organs are absent altogether (*fig.* 20, *b*).

The reproductive bodies of the *Calycophoridæ* are always medusiform, and attached to the peduncles of their respective polypites. In *Vogtia* and *Hippopodius* the manubrium attains a large size, extending far beyond the margin of the short gonocalyx. In other genera the reverse is usually the case, the manubrium being shorter than the swimming cup within which it is suspended. Each gynophore, when fully developed, appears to contain several ova. In most *Calycophoridæ*, except *Diphyes* itself, both male and female reproductive appendages appear on the same hydrosoma.

Four families of *Calycophoridæ* have been defined by Professor Huxley. Their characters we subjoin.

Order CALYCOPHORIDÆ.

Family 1. DIPHYDÆ.
Calycophoridæ with not more than two, polygonal, *nectocalyces*. Proximal *hydrœcium* complete. *Hydrophyllia*.

Family 2. SPHÆRONECTIDÆ.
Calycophoridæ with probably not more than two *nectocalyces;* the proximal one being spheroidal, with a complete *hydrœcium*. No *hydrophyllia?*

Family 3. PRAYIDÆ.
Calycophoridæ with only two *nectocalyces*, whose *hydrœcia* are both incomplete. *Hydrophyllia*.

Family 4. HIPPOPODIIDÆ.
Calycophoridæ with many *nectocalyces*, whose *hydrœcia* are incomplete. No *hydrophyllia*.

The same naturalist has proposed the distinctive term of 'Diphyozoöids' for those singular detached reproductive portions of adult *Calycophoridæ* which received the name of " monogastric *Diphy-*

dæ" from earlier observers. Their true nature was first demonstrated by R. Leuckart, who several times witnessed the separation of these bodies from a well-known species of *Abyla*. Groups of organs became detached from the cœnosarc, each group consisting of a hydrophyllium, polypites, tentacles, and gonophores, with a small portion of the cœnosarc itself. More frequently, however, the actual detachment of the Diphyozoöid has not yet been observed, so that the precise origin of many still presents a subject for inquiry. Pending further investigation, it seems right to designate such forms by provisional generic and specific names, of which not a few have already been conferred.

6. **Order 5: Physophoridæ.**—The *Physophoridæ* differ much more among themselves than do the members of the order just mentioned. All, however, agree in having the proximal end of the cœnosarc modified to form the pneumatophore, or float, which presents so characteristic a feature in the physiognomy of these animals. The cavity of this pneumatophore is a simple enlargement of that of the cœnosarc, the walls of both being directly continuous. To the apex of the cavity is attached a firm, elastic, apparently chitinous sac, known as the 'pneumatocyst,' containing a greater or less proportion of air. A layer of endoderm, reflected from the pneumatophore, invests the whole outer surface of its contained pneumatocyst, which is thus completely cut off from the somatic cavity below. The lower extremity of the pneumatocyst is usually, if not always, entire. Its apex, though most frequently closed, is open in *Physalia* and *Rhizophysa*,

and, the free extremity of the pneumatophore being likewise perforate, a communication exists, in these genera, between the cavity of the pneumatocyst and the surrounding medium. In *Rhizophysa*, moreover, peculiar long branched processes freely depend from the distal surface of the pneumatocyst. Each process consists of a layer of the investing endoderm containing in its axis clear cellæform bodies, ·02 of an inch long, each of which includes an opaque oval endoplast, about $\frac{1}{6}$th of these dimensions, and this, in its turn, a more minute particle or nucleolus, oval or circular in form, and ·0008th of an inch in diameter. In *Agalma* and *Forskalia* radiating membranous partitions connect the walls of the pneumatophore with those of the pneumatocyst, below which each terminates in a free arcuated edge. In *Velella* and *Porpita* the pneumatocyst is furnished with several openings, or stigmata, communicating with the exterior, while to its distal surface are attached a number of long slender processes enclosing air, and hence termed the 'pneumatic filaments.'

Excepting the presence of the pneumatophore and the absence of a somatocyst, the general plan of structure in these *Hydrozoa* differs little from that of the *Calycophoridæ*. In *Apolemia*, as in *Diphyes*, the numerous groups of appendages are supported at intervals along a slender, unbranched, connecting stem. *Physophora*, the type of the order, has a filiform, but comparatively short, cœnosarc, terminated proximally by a pneumatophore of moderate size, below which the greater portion of its length is occupied by a double series of nectocalyces, each alternating with its successor on the opposite side, and deeply grooved on its inner face for attachment to the cœnosarc, (*fig.* 22, *b*). The distal extremity of the latter forms an expanded bulb, above which are disposed, in a spiral or circular manner, the various appendages; consisting of polypites, tentacles, hydrocysts, and organs of reproduction. Of these the hydrocysts are uppermost, or external; next come the polypites, with a tentacle at the base of each, between, or above, which the gonophores, of both sexes, are arranged. The usual length of *Physophora* is about two inches.

The typical genus just described may advantageously be contrasted, on the one hand, with such forms as *Apolemia* or *Halistemma*, on the other, with the widely different, though equally aberrant, genera, *Porpita* and *Velella*.

In *Halistemma rubrum* the appendages are attached to a thread-like stem, nearly forty inches in length, having a float of only three or four lines in its longest diameter, close beneath which the swimming-bells, about sixty in number, extend in two parallel rows for a distance of six or seven inches. The remainder of the cœnosarc is occupied by the polypites, tentacles, hydrocysts, bracts, and reproductive buds, all associated in one continuous series. Especially conspicuous, from their bright vermilion hue, appear the complex tentacular sacculi, while fainter longitudinal bands of the same colour mark the hepatic striæ of the polypites, whose size, in this genus, is considerable. The general aspect of this most beautiful, yet withal, extraordinary being, has been compared by Vogt, its discoverer, to that of a delicate, transparent garland of flowers, endowed, in a marvellous manner, with life and activity.

Far different is the physiognomy of *Velella*, whose cœnosarc appears almost wholly lost in the horizontal, or slightly convex, rhomboidal pneumatophore, which distinguishes this singular genus. The proximal surface of the pneumatophore is traversed diagonally, from one of its angles to the other, by an upright, triangular crest, which, in common with the horizontal disc, consists of a soft

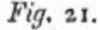

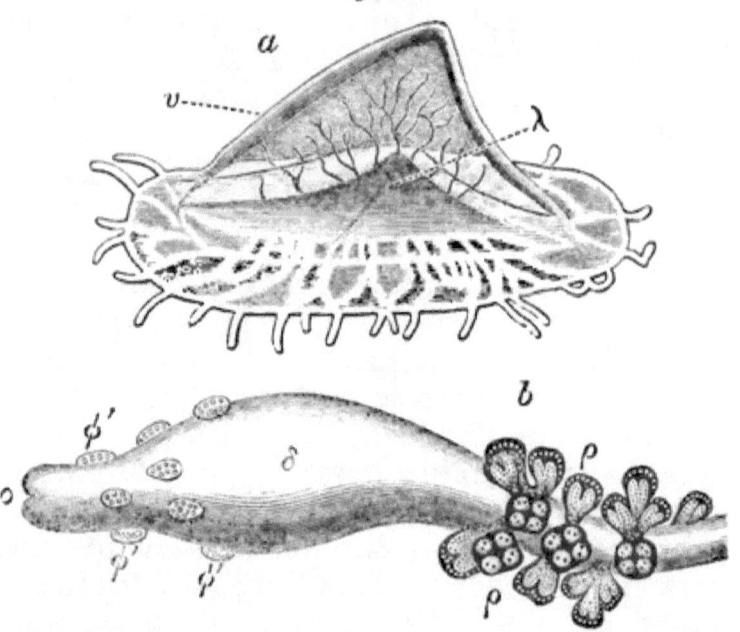

Morphology of VELELLA:—*a*, *Velella spirans*, somewhat enlarged; *b*, one of its smaller polypites, much magnified;—v, crest; λ, liver; o, mouth of polypite; δ, its digestive cavity; ϕ, rounded elevations, containing thread-cells; ρ, medusiform zooids.

marginal membrane, or "limb," bounding the "firm part," or central portion (*fig.* 21, *a*). To the distal surface of the firm part of the disc are attached the several appendages; including (1) a single large polypite, nearly central in position; (2), numerous small gonoblastidia, which resemble polypites, and are termed 'phyogemmaria'; and, (3), the reproductive bodies to which these last give rise (*b*). The tentacles are attached, quite independently of the polypites, in a single series along the line where the firm part and limb of the disc unite. There are no hydrocysts, nectocalyces, or hydrophyllia. The average length of *Velella* may be estimated at two inches, its height at one inch and a half. The entire organism is semi-transparent and tinged with an ultramarine blue, which changes to a deeper shade in the tentacles and limb of the disc.

On closer examination the firm part is seen to enclose a hard, shell-like, pneumatocyst, consisting of a horizontal division, included within the disc, and continuous with the simple solid vertical plate, which gives support to the sail or crest. The upper surface of the pneumatocyst is crossed at right angles to the direction of the crest by a linear diagonal groove, indicated on its under surface by a slightly elevated ridge, " while a longitudinal depression, increasing in depth from the margins to the centre, corresponds with the attachment of the crest. The horizontal division of the pneumatocyst consists of two thin laminæ, passing into one another at their free edges, and united by a number of concentric vertical septa, between which are corresponding chambers filled with air. All these chambers communicate together by means of apertures in the septa. Of these each septum presents two, placed at opposite points of its circumference, and all nearly in the middle line of the pneumatocyst. Kölliker made the interesting discovery that many of the chambers have an additional opening, by which they communicate directly with the exterior. These apertures are situated in the proximal or upper wall of the chambers, along a line about midway between that of the openings just described and that of the vertical plate of the pneumatocyst. Of the thirteen apertures observed by Kölliker, six lay on one side of the vertical plate and seven on the other; one aperture lies in the wall of the central chamber, the other six at tolerably even intervals between this and the margin. Consequently, as there are more than six concentric chambers, some of them must communicate with these stigmata only indirectly." To the under surface of the five or six innermost chambers are attached from ten to fifteen elongated hollow processes containing air, the pneumatic filaments already mentioned.

But complicated as the pneumatocyst of *Velella* may seem, not less so is its curiously modified somatic cavity. On all sides the limb is traversed by an anastomosing system of canals, which are ciliated, and communicate with the cavities of the phyogemmaria and large central polypite. Within the roof of the latter, close beneath the pneumatocyst, is lodged a peculiar brownish mass, the so-called liver. This, also, is furnished with a canal system of its own, which eventually becomes continuous with the sinuses of the limb.

In addition to the preceding organs *Velella* possesses certain large "glandular sacs," for the discovery of which we are indebted to Vogt. He describes them as presenting a very curious minute structure, and as arranged in a single series around the margin of the limb, to open on its dorsal surface, where they secrete a clear, viscid, mucus. The true nature of this mucus, whether excretory or lubricative, is still very imperfectly known.

Thus the most striking modifications of the common plan of the *Physophoridæ* depend on differences in the relative size and shape of the cœnosarc and pneumatophore. *Athorybia* and *Physalia* have, like *Velella* and *Porpita*, a disproportionately large pneumatophore; but, in these genera, it is globular or pear-shaped, not, as in those, discoidal. In *Physalia*, the true Portuguese Man-of-war of sailors, often wrongly regarded as the type of the present group, the float sometimes attains a long diameter of eight or nine inches; tentacles, several feet in length, being attached directly to the cœnosarc along its under surface (*fig.* 11, *c*). But, more frequently, the cœnosarc is filiform, with a small pneumatophore; and, except in the case of *Rhizophysa*, swimming-bells are also present. Nectocalyces and hydrophyllia are alike absent in *Porpita, Velella, Physalia*, and *Rhizophysa*. *Athorybia* has hydrophyllia, without nectocalyces; *Physophora* nectocalyces, but no hydrophyllia. All other genera possess these two kinds of appendages.

The swimming-bells of the *Physophoridæ*, when present, are more numerous than in the *Calycophoridæ*, and, among the different genera, vary much in size, shape, and mode of attachment, as also in the relative proportions of the nectosac. Each frequently has its surface marked with grooves and ridges, and may send forth processes which serve to embrace the cœnosarc, and connect it with its fellow of the opposite side. In some genera, more particularly *Physophora* itself, two of the nectocalycine canals, which coincide with what may be termed the medial plane of the cœnosarc, remain, as usual, straight, while the two other, or lateral, vessels become convoluted in a most complicated manner before reaching the circular canal. As in the *Calycophoridæ*, the common cavity of each nectocalyx is connected with that of the cœnosarc by means of a tubular pedicle.

The hydrophyllia present variations both in their structure, mode of attachment, and relations to the other appendages. They may be either foliaceous (*Athorybia*), or clavate (*Apolemia*), or thick and wedge-like, or even pyramidal (*Agalma*); while their surface is liable to be diversified with excavations and ridges, having smooth or serrate lateral margins. Their arrangement is, in general, more or less whorled. In *Apolemia* they are proximal to all the other appendages, in each separate group, and here, as in *Halistemma* and *Stephanomia*, they become connected with the cœnosarc by more or less distinct peduncles. In *Agalma*, the attached apex of each is pierced by a duct which terminates in a cæcal phyllocyst, about the middle of the hydrophyllium, while its opposite end opens into the somatic cavity. In *Forskalia* the hydrophyllia are attached directly to the peduncles of the polypites. The graceful *Athorybia rosacea* possesses from twenty to forty of these organs, inserted, in two or three circlets, immediately below the pneumatocyst, and above a much smaller number of polypites (*fig.* 22, *a*). In all other *Physophoridæ* with hydrophyllia, nectocalyces also are present; but *Athorybia*, though destitute of the latter appendages, has the power of alternately raising and depressing its hydrophyllia, so as to render them agents of propulsion.

Fig. 22.

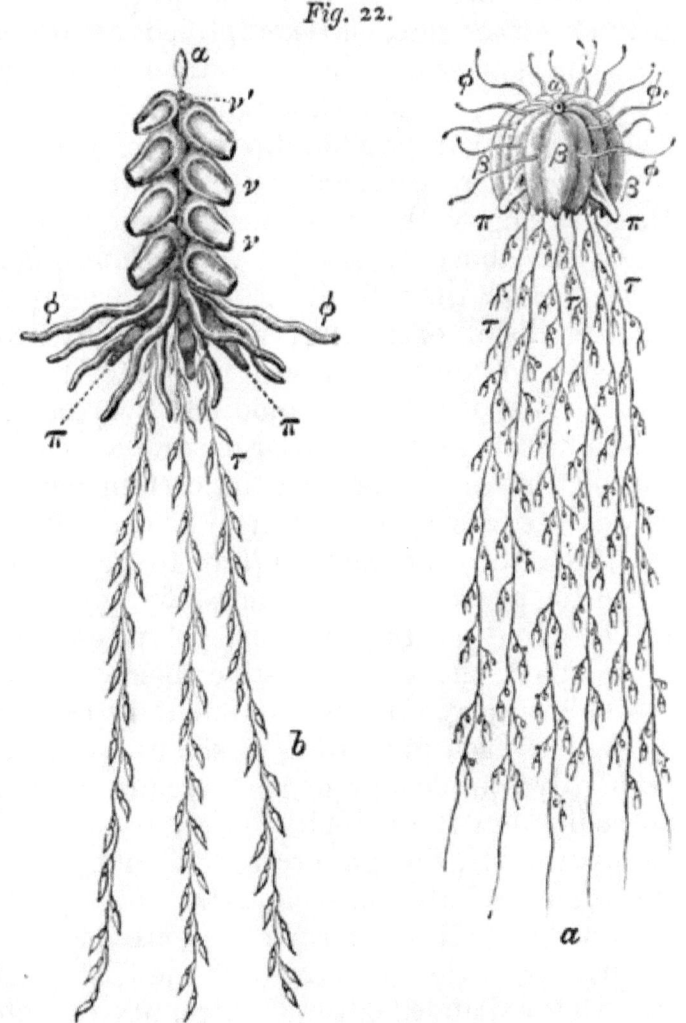

Morphology of PHYSOPHORIDÆ:— *a*, *Athorybia rosacea*; *b*, *Physophora Philippii*; α, pneumatophore; ν, nectocalyx; ν′, rudimentary nectocalyx; φ, hydrocysts; π, polypite; τ, tentacles; β, hydrophyllia. (About the natural size.)

The polypites of the several genera differ chiefly in size and mode of attachment. They may be inserted in continuous series, along one side of the cœnosarc only, as in *Stephanomia*, or indifferently on either side, as exemplified by *Rhizophysa*. In some cases they are attached directly to the cœnosarc; in others, supported, with their tentacles and hydrophyllia, upon special stalks. In *Apolemia* they are arranged in groups of two or three, along with the other appendages, at intervals, as above mentioned; in *Physophora* and *Athorybia* they form a spiral or circlet around its distal extremity, while in *Physalia*, *Velella* and *Porpita*, they are restricted to the inferior surface of the much modified hydrosoma.

Two kinds of polypites, a larger and a smaller, appear on the same cœnosarc in certain genera. Much doubt exists as to the true nature of the latter, which in some cases appear to have been mistaken for hydrocysts, in others, for gonoblastidia. In *Agalma* the smaller polypites often equal in length the true digestive zoöids, but are always much narrower, the rudimentary tentacle at the base of each presenting a striking contrast to the highly complex prehensile organs attached to the pedicles of the polypites properly so called.

The hydrocysts, though present in many other genera, are especially conspicuous in *Physophora* and *Athorybia*. In the former they are disposed, in a circular series, external to the polypites, around the expanded distal extremity of the cœnosarc, and, from their bright pink colour and larger size, are more noticeable, on a first inspection, than the true digestive appendages. (*fig.* 22, *b.*)

Attention has, in a previous section, been directed to some of the modifications which the tentacles of the *Physophoridæ* present. They appear in *Apolemia* as simple tubular processes, with numerous large thread-cells on one side: in *Velella*, they are equally simple, but much shorter, and slightly enlarged at their free ends. In *Porpita* there is, in addition, a series of longer prehensile appendages, having their distal extremities clavate and beset with stalked knobs, or capitula, containing urticating organs, which are wanting in the smaller marginal "cirrhi." In *Physalia*, as above described, each tentacle consists of a broad conical basal sac, and a long simple ribbon-like process, having transverse reniform enlargements (*fig.* 11, *d*). In all other *Physophoridæ* the tentacula are furnished with lateral branches, which in *Rhizophysa* alone appear to be destitute of sacculi. These may want involucra, as in *Halistemma* or *Forskalia*, or possess these structures, and become further modified, as in all the remaining genera.

The structure of the gonophores, though always medusoidal, is, in other respects, liable to much variation. Among many *Physophoridæ* well-marked differences of size, aspect, or relative number, distinguish the male and female reproductive bodies in the same species. Thus in *Agalma* and *Physophora* the gynophores are only half the length of the androphores, than which, however, they are more numerous. In *Velella* and *Porpita* both androphores and gynophores become completely detached, and the same is probably true of the female organs in *Physalia*. But the androphores of this genus, and the two kinds of generative appendages in many other forms, discharge, in all probability, their proper functions, without previous separation from the parent hydrosoma. The gonophore either presents a well-developed swimming-cup, with open margin and conspicuous canals: or the calyx, with its canal system, may remain in a very rudimentary condition, so as scarcely to be distinguishable from its contained manubrium; while in the male organs of *Stephanomia* and *Athorybia* the apex of the latter slightly projects beyond the margin of the bell. In these and many other genera each gynophore gives rise to only a single ovum; but there is reason to infer that it may be otherwise with the free-swimming gonophores of *Physalia* or *Velella*.

Most of the *Physophoridæ* hitherto examined appear to be monœcious, the androphores and gynophores being borne on the same gonoblastidia, as in *Physalia*, *Agalma*, and *Athorybia*, or, more rarely, on separate stalks, as in *Stephanomia*. Besides the reproductive bodies, the gonoblastidia may give support to hydrocysts and other appendages, as in *Physalia* or *Athorybia*. In *Halistemma* they are absent or so extremely short, that the gonophores seem attached directly to the cœnosarc. The phyogemmaria of *Velella* and *Porpita* are obviously homologous with the gonoblastidia of *Hydractinia* or the small polypites of *Podocoryne*; and just as the former genera differ from other *Physophoridæ*, *Physalia* indeed excepted, in their laterally expanded cœnosarc and independent tentacles, so, likewise, may *Hydractinia* be distinguished from all the more typical forms of *Corynidæ*.

Of *Physophoridæ* Mr. Huxley has established

seven definable families, whose characters may be stated thus:—

Order PHYSOPHORIDÆ.

Family 1. APOLEMIADÆ.
Pneumatocyst small. *Cœnosarc* filiform.
Nectocalyces and *hydrophyllia* present; the latter united with the other appendages into groups, arranged at distant intervals along the cœnosarc.
Tentacula without lateral branches.

Family 2. STEPHANOMIADÆ.
Pneumatocyst small. *Cœnosarc* filiform.
Nectocalyces and *hydrophyllia* present; the latter arranged with the other appendages in continuous series.
Tentacula with lateral branches, terminated by sacculi.

Family 3. PHYSOPHORIADÆ.
Pneumatocyst small. *Cœnosarc* filiform, but short, and dilated at its distal end.
Nectocalyces present, occupying most of the cœnosarc below the pneumatophore. No *hydrophyllia*.
Tentacula with branches and involucrate sacculi.

Family 4. ATHORYBIADÆ.
Pneumatocyst occupying almost the whole of the globular *cœnosarc*.
Nectocalyces absent. *Hydrophyllia*.
Tentacula with branches and involucrate sacculi, each having two filaments and a median lobe.

Family 5. RHIZOPHYSIADÆ.
Pneumatocyst small. *Cœnosarc* filiform.
Nectocalyces and *hydrophyllia* absent.
Tentacula having branches, without sacculi.

Family 6. PHYSALIADÆ.
Pneumatocyst occupying almost the whole of the thick and irregularly fusiform *cœnosarc*.
Nectocalyces and *hydrophyllia* absent.
Tentacula with basal sacs, but no lateral branches.

Family 7. VELELLIDÆ.
Pneumatocyst flattened, divided into chambers by numerous concentric partitions, and occupying almost the whole of the discoidal *cœnosarc*.
Nectocalyces and *hydrophyllia* absent.
Tentacula short, clavate, simple or branched, submarginate.

A single central, principal polypite.

7. **Order 6: Medusidæ.**—If the phyogemmaria of *Velella* and *Porpita* be regarded as gonoblastidia, the hydrosoma of these genera may then be said to present not more than one true polypite; a character in which they differ from all other *Physophoridæ*, but agree with the members of the next order, *Medusidæ*. Cuvier, indeed, associated the *Velellidæ*, *Medusidæ*, and free zoöids of the *Lucernaridæ* in a single group, under the name of " Acalèphes Simples"; the remaining *Physophoridæ*, together with the *Calycophoridæ*, being distinguished as " Acalèphes Hydrostatiques". But the *Velellidæ*, like all other

Physophoridæ, differ, as has been shown, from *Calycophoridæ*, in possessing a float, and from *Medusidæ* in the characteristic mode by which the polypite is connected with the rest of the hydrosoma. In both the *Calycophoridæ* and *Physophoridæ*, the nectocalyces (when present) and polypites are separately attached to different parts of the cœnosarc. In the *Medusidæ*, on the other hand, the hydrosoma presents but one nectocalyx, from the roof of which a single polypite is suspended (*fig.* 4, *f*). The endodermal lining of the polypite passes into the central cavity of the swimming-organ, from which, as in other nectocalyces, canals radiate, to join a circular vessel surrounding the margin of the bell. From this margin depend tentacles, which may be either hollow processes of both layers, in immediate connection with the canal system, or, more rarely, prolongations of the gelatinous ectoderm itself. Around the outer margin of the nectocalyx, between the endoderm of the circular vessel and its ectodermal investment, are embedded the marginal bodies, vesicles or pigment-spots, whose peculiar structure has already been described (*fig.* 23).

The outward form of the polypite varies greatly. It may be long and highly contractile, or stoutly cylindrical, or so short and broad as to be with difficulty discernible on the under surface of the bell (*fig.* 24). Very often it is curiously constricted. In internal structure it is not known to present any peculiar features. The oral margin may be either simple, everted, or produced into lobes, which, most frequently, are four in number, though in some forms it is much divided. In *Liriope*

Catharinensis, it is surrounded by a series of little sacs, each well packed with thread-cells.

The size and shape of the nectocalyx in relation to the polypite with which it is connected may also vary considerably. The veil which surrounds the open margin of the nectosac in no case appears to

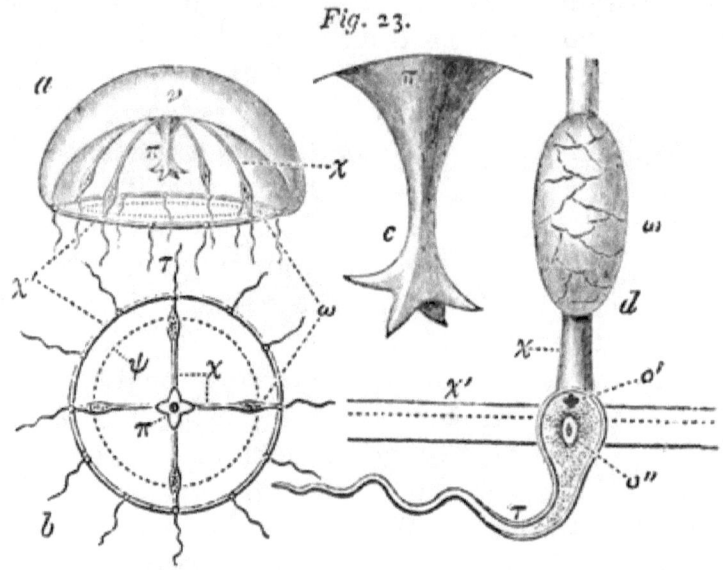

Fig. 23.

Morphology of MEDUSIDÆ:—*a*, Medusid, seen in profile; *b*, the same, viewed from below; *c*, its polypite; *d*, part of its marginal canal, and other structures in connection therewith;—ν, nectocalyx; π, polypite; χ', marginal canal; ψ, veil; τ, tentacle; χ, radiating canal; ω, reproductive organ; o', coloured spot; o'', marginal vesicle.

be absent. More than four longitudinal canals sometimes occur. In *Willsia* these canals are seen to bifurcate, each branch again dividing into two others, so that, in this form, the six canals open by twenty-four ducts into the circular vessel, (*fig.* 24, *c*). In *Cunina, Ægina*, and *Æginopsis*, both circular and radiating canals disappear, their place being supplied by peculiar pouch-shaped processes, communicating with the digestive cavity of the polypite.

The tentacles scarcely require any special mention. Usually the distal extremities of their cavities

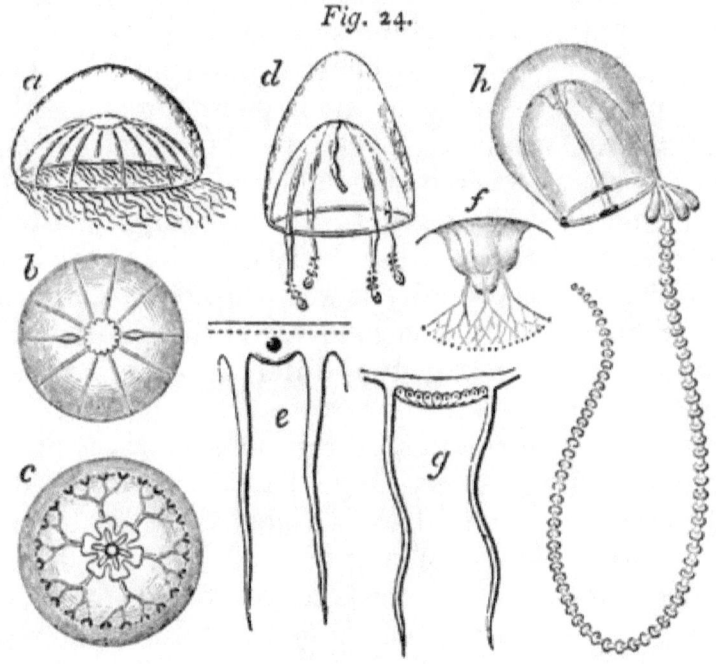

Fig. 24.

Various forms of MEDUSIDÆ:—*a*, *Æquorea formosa*, seen in profile; *b*, the same, viewed from above; *c*, upper view of *Willsia stellata*; *d*, *Slabberia conica*; *e*, portion of the marginal canal of *Tiaropsis Pattersonii*; *f*, polypite of *Bougainvillea dinema*; *g*, part of its marginal canal; *h*, *Steenstrupia Owenii*. (*a*, *b* and *d* are about the natural size; the others are magnified.)

become more or less obliterated through vacuolation of the endodermal lining. In *Trachynema* and its allies the tentacles are stiff; not contractile, as in other *Medusidæ*.

The reproductive organs have been stated to be of the simplest kind, consisting of mere expansions, either of the polypite wall, or radiating canals, within which the generative elements are produced.

At a time when the free gonophores of the *Hydrozoa* had been as yet imperfectly studied, it was the custom of naturalists to regard these bodies as independent individuals, worthy of being arranged under definable genera and species. The singular resemblance of such gonophores to the *Medusidæ* began, at length, to attract attention. Then it was suspected that many of the *Medusidæ* were not individual organisms properly so called, but merely the free reproductive buds of various *Hydrozoa*. Eventually it was proposed to abolish the whole group of *Medusidæ*, and distribute their several forms among the different orders of the class.

On the other hand, certain observations of J. Müller, Fritz Müller, Gegenbaur, and Claparède, to which we have already referred, indicate the probable existence of a group of Medusid forms which appear to be the immediate products of true generative acts, not of gemmation or fission, (*fig.* 12).

In the present state of our knowledge, it seems better to sum up the several aspects of this doubtful question in the following series of conclusions.

1. That several of the organisms formerly described as *Medusidæ* are the free gonophores of other orders of *Hydrozoa*.

2. That the homology of these free gonophores with those simple expansions of the body-wall which in *Hydra* and some other genera are known to be reproductive organs by their contents alone

is proved alike by the existence of numerous transitional forms, and an appeal to the phenomena of their development.

3. That many other so-called *Medusidæ* may, from analogy, be regarded as, in like manner, medusiform gonophores.

4. But that there may exist, nevertheless, a group of Medusid forms, which may give rise, by true reproduction, to organisms directly resembling their parents, and, therefore, worthy of being placed in a separate order under the name of *Medusidæ*.

All the *Trachynemidæ* and *Æginidæ* belong, according to Gegenbaur, to the order in question. And to the same group may be referred, provisionally, that large assemblage of forms anatomically similar to true *Medusidæ*, but whose development is unknown; just in the same manner as genera and species are established for those Diphyozoöids which, there is every reason to believe, are but the detached fragments of other *Calycophoridæ*. Pending the study of the life-history of these ambiguous Medusoids, their true nature must, also, remain undetermined.

Such are the forms brought together by Gegenbaur in the systematic table here annexed, which, at the same time, concisely displays their most striking anatomical peculiarities.

Order MEDUSIDÆ

With radiating canals.
    Reproductive organs in the polypite-wall. Ocelli at bases of the tentacles . . . Family 1. OCEANIDÆ.

Reproductive organs in the course of the radiating canals.
  Radiating canals arising from the base of the polypite. Ocelli. . . . . Family 2. THAUMANTIADÆ.
  Radiating canals arising from the outer margin of the polypite. Vesicles. . . . Family 3. ÆQUORIDÆ.
Reproductive organs as rounded protuberances of the radiating canals. Vesicles.
  Tentacles contractile. . . Family 4. EUCOPIDÆ.
  Tentacles stiff. . . . Family 5. TRACHYNEMIDÆ.
Reproductive organs as flattened expansions of the radiating canals. Vesicles. . . . Family 6. GERYONIDÆ.
With pouch-shaped prolongations of the polypite, in which the reproductive products are formed. Vesicles. . . . . . Family 7. ÆGINIDÆ.

The *Medusidæ* were termed by Eschscholtz, Cryptocarpæ; by E. Forbes, Gymnophthalmata; and by Gegenbaur, Craspedota. These words contrast, respectively, with the names Phanerocarpæ, Steganophthalmata, and Acraspeda applied by the same naturalists to a large section of the *Lucernaridæ*. In this group the family *Lucernariadæ* is not usually included, many naturalists, from a mistaken view of its organisation, referring it to the class *Actinozoa*, of which Professor Milne Edwards has recently made it a distinct sub-class, under the title of Podactinaria.

8. **Order 7: Lucernaridæ.**—In these *Lucernariadæ* the body is more or less cup-shaped, and frequently about an inch in height, terminating proximally in a stalk of variable length, and furnished with a hydrorhiza, which, like that of

Hydra, is not permanently attached. Round the distal margin of the cup arise a number of short tentacles which, in *Lucernaria* itself, are disposed in eight or nine tufts, but in *Carduella* form one continuous series. Their free extremities appear sucker-like or capitate; in *Depastrum*, however, they are simply clavate. The whole organism is semi-transparent, variously coloured, and of a gelatinous consistence (*fig.* 25).

The cup, in Carduella, presents at its centre a four-lobed mouth, which is easily seen to form the free extremity of a distinct polypite, occupying the axis of the entire hydrosoma. The oral margin of this polypite is simple and slightly everted. Its gastric region exhibits a number of tubular filaments, arranged in vertical rows, and projecting freely into the digestive cavity. In transverse section the polypite may be described as somewhat quadrilateral, with a sinuous outline, which expands at its four angles to form as many deep longitudinal folds, within which the simple generative bands are lodged. The space between the polypite-wall and the inner surface of the cup is divided in the following manner. From each projecting angle of the gastric region run a pair of vertical septa, which diverge widely from one another so as to reach the wall of the cup at points precisely opposite the two sinuosities on either side of the generative band. Thus four of the equidistant lines along the inner surface of the cup receive two converging septa, each, however, belonging to a different pair. These last septa, with the polypite wall, serve to enclose four wide longitudinal canals, outside of which are four other spaces, bounded, within, by two septa of the same pair, and, externally, by the cup itself. The outer canals are closed superiorly by a roof, consisting of four inflected lobes from the summit of the cup; the inner spaces remaining open. There are also four very narrow canals coinciding with the lines where the vertical septa and inner surface of the cup meet; and formed, it would seem, partly by these septa and partly by folds in the cup's substance. Lastly a circular sinus has its course immediately beneath the insertion of the tentacles.

Fig. 25.

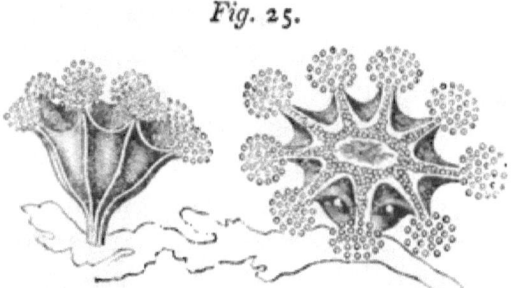

LUCERNARIA:—Two specimens of *Lucernaria auricula*, attached to a piece of sea-weed. That figured to the right is somewhat abnormal, having a ninth tuft of marginal tentacles. (Natural size.)

By means of a band of muscular tissue which traverses its margin, and another set of fibres which radiate towards the polypite, the distal extremity of the cup can fold inwards and contract itself at the pleasure of the animal. Some *Lucernariadæ* have been observed to detach themselves, and swim in an inverted position by the slowly repeated movements of their cup-like umbrella. In this respect they agree with *Pelagia*, a much more active and permanently free member of the same order.

The swimming organ of *Pelagia* is sub-globose, about three inches in diameter, divided at its margin into sixteen lobes. Under eight of these lobes are seen notches, each lodging a hooded lithocyst, while from the remaining lobes depend an equal number of long, contractile tentacula. A polypite, short and broad, is attached, proximally, to the concave centre of the umbrella; distally, it terminates in four furbelowed lips, which extend to a length of nearly four inches. A number of cæcal sacs, corresponding with the lobes of the umbrella, are prolonged from the digestive cavity. In other characters *Pelagia* resembles the free zoöids of *Aurelia* and its allies (*fig. 7, b*).

The *Lucernaridæ* admit of being arranged under two principal sections, in one of which the development is continuous, in the other, discontinuous. The first section includes *Pelagia* and the *Lucernariadæ*, in which reproductive elements are produced by the organism immediately resulting from a generative act. In other members of the order, the primitive result of this act is a fixed and sexless 'Lucernaroid,' which gives rise by fission to free zoöids of disproportionate size, in which the reproductive organs are developed. The first section, again, includes two minor divisions, in one of which the umbrella is permanently free, in the other, furnished with an organ of attachment. But the developmental cycle of each Lucernarid belonging to the second section presents these two principal forms.

It is worthy of remark, that the Lucernaroids of very different genera — such as *Cephea* and *Chrysaora* — are often wonderfully alike in structure; so that the relation between the producing and produced zoöid is here by no means the same as in the other orders of *Hydrozoa*. The true import of this fact should not escape attention.

All the *Lucernaridæ* may be at once distinguished by their umbrella. The cup or disc in the *Lucernariadæ* and Hydra-tubæ, the swimming organ of *Pelagia* and of the free zoöids, are alike included under this designation. A free umbrella differs from a nectocalyx, with which it is often confounded, (1), in the absence of a veil; (2), in its mode of development; and (3), in the nature of its canal system and marginal bodies. The radiating canals, never less than eight in number, send off numerous anastomosing branches, which form a very intricate net-work. The peculiar structure of the lithocysts has been previously explained. Each is supported on the end of a short double-walled stalk, the cavity of which runs into one of the radiating canals. Protection is given to this apparatus by a hood-like, crescentic fold of the ectoderm, at the base of which, and on the convex surface of the umbrella, a funnel-shaped orifice has been observed, whereby the radiating canal communicates with the exterior. Apertures similar in function, but not in position, have been met with by Mr. Huxley in certain of the *Medusidæ*. There are no lithocysts in the *Lucernariadæ*, unless the simple tubercles, placed between the tentacular tufts, on the margin of *Lucernaria auricula*, be regarded as these organs in a rudimentary condition.

The *Lucernaridæ* manifest another characteristic feature in their gastric filaments, the presence of which appears to be universal throughout the order. They may, without difficulty, be observed in the common species of *Lucernaria*. They are usually solid, perhaps through vacuolation; contain thread-cells, and, even when detached, execute a peculiar writhing kind of movement. In *Chrysaora*, according to Fritz Müller, they attain a length of several inches.

Three families of *Lucernaridæ* may be defined as follows:—

Order LUCERNARIDÆ.

Family 1. LUCERNARIADÆ.
Reproductive elements developed in the primitive hydrosoma, without intervention of free zoöids.

Umbrella with short marginal tentacles and a proximal hydrorhiza.

Polypite single.

Family 2. PELAGIDÆ.
Reproductive elements developed in a free umbrella, which either constitutes the primitive hydrosoma or is produced by fission from an attached Lucernaroid.

Umbrella with marginal tentacles.

Polypite single.

Family 3. RHIZOSTOMIDÆ.
Reproductive elements developed in free zoöids produced by fission from attached Lucernaroids.

Umbrella without marginal tentacles.

Polypites numerous, modified, forming with the genitalia a dendriform mass depending from the umbrella.

Not far from *Pelagia*, but in a family by itself, Gegenbaur has placed the genus *Charybdea*. Fritz Müller, however, shows, that in the closely allied *Tamoya*, a distinct veil is certainly present, while *Charybdea* itself is furnished with marginal processes, which seem to represent the same apparatus.

Section IV.

DISTRIBUTION OF HYDROZOA.

1. Relations to Physical Elements. — 2. Bathymetrical Distribution. — 3. Geographical Distribution.

1. Relations to Physical Elements.— The *Hydrozoa*, as a class, are almost exclusively marine; *Hydra* and *Cordylophora* being the only fresh-water genera hitherto described.

2. Bathymetrical Distribution.— The marine *Hydrozoa*, with reference to their distribution, may conveniently be divided into two groups, the fixed and the oceanic. The fixed *Hydrozoa*, *Corynidæ* and *Sertularidæ*, are less abundant between tide marks than at depths of a few fathoms, some forms extending their range to very deep water. The *Corynidæ* are, perhaps, on the whole, more partial to shallow waters than the *Sertularidæ*, certain species of the latter order, especially of the genus *Campanularia*, being found at considerable depths. But the vertical distribution of several forms is more limited than that of others. Thus *Clava* and *Coryne* appear usually not to wander far from low-water mark, while *Tubularia* occurs at depths varying from less than one to more than fifty fathoms.

The oceanic *Hydrozoa* in fine weather swim near the surface of the water, the approach of rain or wind compelling them to retire for safety to the more tranquil depths below. The large "jelly-fishes" which, during summer and autumn, occur so abundantly in our seas, are, with few exceptions, the reproductive zoöids of *Aurelia*, *Cyanea*, and *Chrysaora*. Equally numerous with these, but less conspicuous from their extreme transparency, appear hosts of minute medusoids, while Diphyozoöids, *Velella*, *Physalia*, and one or two other *Physophoridæ* may, at rarer intervals, be detected.

3. **Geographical Distribution.**—The genera of *Hydrozoa* are very widely distributed, renewed investigations tending rapidly to diminish the number of those supposed to be peculiar to certain regions of the globe.

The limits of the area inhabited by *Hydra* have not yet been definitely ascertained. The other fresh-water genus, *Cordylophora*, has been met with only in Denmark, Great Britain, Ireland and North America.

Not much is known accurately of the geographical range of the *Corynidæ*; the *Sertularidæ*, from the ready preservability of their polypary, having been far more extensively studied. *Sertularia*, *Plumularia*, *Antennularia*, and *Campanularia* are truly cosmopolitan, and the same may, likewise, be said of some species of these genera, for example, *S. operculata*. Many South African

Sertularidæ are identical with European forms, both being, in a large number of cases, sufficiently distinct from the Australian, Phillipine, and New Zealand species. Several British species cannot, however, be distinguished from those of the Atlantic coasts of America, while on the other hand, greater differences prevail between these last and the North Pacific forms. Among purely exotic genera may be mentioned *Cryptolaria*, one species of which has been found in Madeira and another in New Zealand, and *Lineolaria*, a remarkable Australian Sertularid, having gonophores with two longitudinal rows of strong spines elevated in ridges, between which a few smaller spines are scattered over a flattened, transversely furrowed area.

The *Calycophoridæ* and *Physophoridæ* have hitherto been most successfully studied in the Mediterranean and Southern seas; some genera, such as *Diphyes* and *Agalma*, having been obtained by Sars at a latitude of $61\frac{1}{2}°$ N. off the shores of Norway. It were premature to describe any of these forms as peculiar to certain regions, many of the species and genera ranging over areas of considerable magnitude.

Precise information is much wanting on the distribution of the *Medusidæ* and *Lucernaridæ*. The free zoöids of some species are very extensively diffused, and are occasionally met with by sailors in numbers so immense as almost to impede navigation. Our common *Aurelia aurita* has been obtained in the Red Sea, off the east coast of North America, and in various parts of the southern hemisphere.

A few words on the phosphorescence of the

Hydrozoa may here be inserted. This property has been observed in most orders of the class, though, among the *Physophoridæ, Stephanomia,* and, of the *Lucernaridæ, Pelagia* are most remarkable for its manifestation. Some, however, of our more common jelly-fishes are also luminous. It does not appear that, in any of these, special light-giving organs exist. In the *Medusidæ,* the phosphorescence chiefly arises from around the marginal bodies, but, in some instances, it is emitted by the reproductive swellings, and, occasionally, by the walls of the central polypite. Our own *Thaumantias lucifera,* a species by no means rare, displays this phenomenon in a very beautiful manner. The little creature, when irritated by contact of fresh-water, marks its position by a vivid circlet of tiny stars, each shining from the base of a tentacle.

Such small *Medusidæ* are, doubtless, more efficient in promoting the luminosity of the ocean than their larger and, at times, more brilliantly conspicuous congeners. But the fixed *Hydrozoa,* which, obviously, can take no share in this display, are, also, eminently phosphorescent. A remarkable greenish light, like that of burning silver, may be seen to glow from many of our native *Sertularidæ,* becoming much brighter under various modes of excitation. It is an error to suppose, however, that thus alone do these cold, oily, flames emanate. "If (writes Professor E. Forbes) a bunch of one of the bushy corallines, such as *Sertularia abietina,* be plunged when active and alive imto fresh-water or spirits, a gorgeous display of living stars is instantaneously produced."

SECTION V.

RELATIONS OF HYDROZOA TO TIME.

Well-preserved remains of extinct *Hydrozoa* are wanting. Obscure indications of fossil *Sertularidæ* and, perhaps, also of *Lucernaridæ*, have on a few occasions been met with, but of too fragmentary a character to permit of definition. Professor Agassiz, indeed, states that, many years ago his " attention was attracted by two slabs of limestone slate from Solenhofen, the counterparts of one another, upon which a perfect impression of a Discophorous Acaleph was distinctly visible."

The Graptolites and Oldhamiæ have, by some naturalists, been referred to the present group. Both, however, may, with more propriety, find a place in the Molluscan class of *Polyzoa*.

CHAPTER III.

THE CLASS ACTINOZOA.

SECTION I.

MORPHOLOGY AND PHYSIOLOGY OF ACTINOZOA.

1. Type of the Class: Actinia. — 2. General Morphology. — 3. Organs of Nutrition. — 4. Prehensile apparatus. — 5. Tegumentary Organs. — 6. Corallum or Skeleton. — 7. Muscular System and Organs of Locomotion. — 8. Nervous System and Organs of Sense. — 9. Reproductive Organs.

1. **Type of the Class: Actinia.** — The *Actinia*, or Sea-anemone, is the type of the class *Actinozoa* (*fig.* 26, *c*).

The body of *Actinia* presents a soft, fleshy, or leathery consistence, and varies much both in form and size, according as it assumes its contracted or expanded condition. Average specimens attain, when expanded, a diameter of from one to three inches, their height being rather less; but these dimensions are often exceeded. The expanded *Actinia* is somewhat cylindrical in figure, attaching itself by one of its flattened ends, known as the 'base,' a mouth being placed in the centre of the 'disc,' or opposite extremity. Numerous tentacles, disposed in alternate series, surround the disc's outer margin, between which and the mouth a region destitute of any appendages, the 'peristomial space,' is usually observable.

The mouth is of a slightly elliptical form, a pair of tubercles, including between them a groove, being situated at each of two opposite points of its circumference. The stomach, or digestive sac, into which the mouth directly leads, is a short, distensible tube, open at both ends, and extending about half-way towards the base of the animal; in diameter scarcely exceeding the mouth itself, with which its form, when viewed from above, is seen to correspond. The folds, or grooves, between the oral tubercles are continued, in the form of semi-canals, along the inner surface of this stomach, until, finally, they reach the wide aperture by which it communicates with the somatic cavity.

A transverse section of the body of *Actinia* exhibits two concentric tubes, the outer being constituted by the body-wall, the inner by the digestive sac. The wide space which intervenes between these tubes is divided by a number of radiating partitions, or 'mesenteries,' arising at definite intervals from the inner surface of the body-wall. The 'primary,' or first-formed and widest, mesenteries serve to fix the stomach in its place, their inner edges being inserted throughout the entire length of its outer surface. From the base of the stomach, the inner edge of each mesentery, becoming free, arches, at first, abruptly outwards, and then, more gradually, downwards and inwards, until at length it reaches the centre of the base, from which all the primary mesenteries appear to radiate. Other partitions, developed in successive cycles between those just mentioned, and having no connection with the stomach wall, are distinguished, in accordance with their relative narrowness, as 'secondary mesenteries,' 'tertiary mesenteries,' and so on.

Fig. 26.

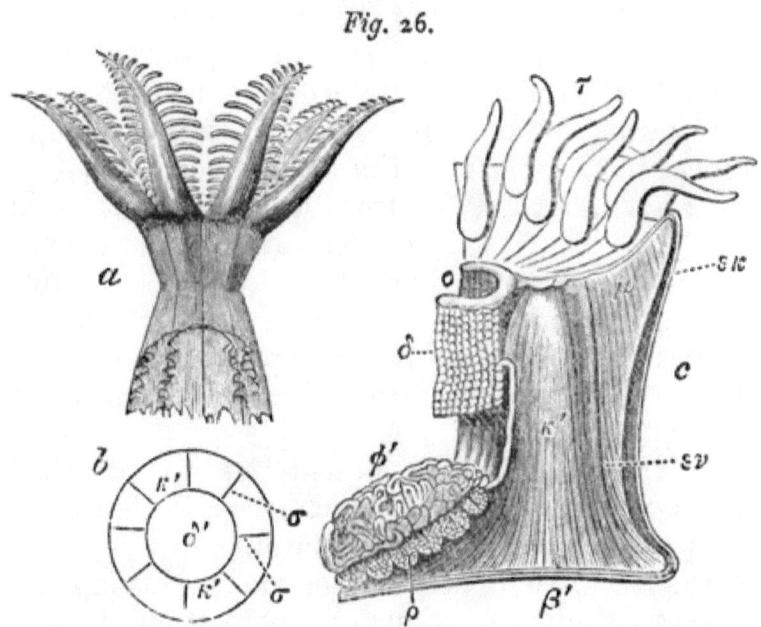

Morphology of ACTINOZOA:—*a*, polype of *Alcyonium;* *b*, ideal transverse section of the same; *c*, longitudinal section of *Actinia;* —κ', somatic cavity; σ, mesentery; δ', digestive cavity; δ, wall of digestive cavity; o, mouth; τ, tentacles; εκ, ectoderm; εν, endoderm; μ, muscular layer; β', base; ρ, reproductive organs; φ', convoluted filaments, containing thread-cells. (*a* and *b* are enlarged; *c* is of the natural size.)

The mesenteries of each cycle are arranged in alternate pairs, while those belonging to opposite sides of the body correspond and are similar to one another. Externally, the mesenteries are often indicated by lines or ridges which traverse the whole length of the column, and are continued, in radii, along the base and disc. Their arrangement is best seen in living, semi-transparent species, without any recourse to dissection.

Thus, a number of imperfect chambers are formed, all opening into one another below, or beyond the free edges of the mesenteries; and, in some cases, apertures occur in the sides of the mesenteries themselves by which a further communication is kept up. These apertures usually appear in the midst of the wide upper portion of the mesentery, not far from the under surface of the disc. They are most constant in the primary partitions; the secondary mesenteries being frequently imperforate. The tentacles, which are hollow, and, in many Actiniæ, perforate at their free extremities, open directly into the somatic chambers.

To the faces of the mesenteries are attached the reproductive organs, which occur as thickened bands of a reddish tint, containing ova or spermatozoa. The male and female organs appear perfectly similar, previous to examination of their contents. Most Actiniæ are diœcious, but, by no external character can the individuals of both sexes, which seem to be about equally numerous, be distinguished from each other. Accurate observations are yet wanting on the reproduction of *Actinia*. It is probable that the spermatozoa, first diffused in sea-water, find their way through the mouth to the ova contained in the general cavity of the body.

A long convoluted cord, or 'craspedum,' arises in front of the reproductive apparatus, along the free edge of each mesentery. In addition to the craspeda, other organs of similar structure, termed 'acontia,' are occasionally met with. Both craspeda and acontia are richly furnished with thread-cells, for the emission of which special apertures along the wall of the somatic cavity have, in some species, been observed. Mr. Gosse, who gives the name of 'cinclides' to these apertures, describes them as varying considerably in size and opening directly into the somatic chambers. "Each is an oval depression, with a transverse slit across the middle." The sides of the cinclis can be opened or closed at the animal's pleasure, yet, when separated to their utmost extent, the front of the orifice is seen to be protected by a very thin superficial film.

In the common Sea-anemone the margin of the disc is furnished with a series of white or bright blue specks, which some writers describe as a rudimentary apparatus of vision. The structure of these organs is not yet fully understood. Like the body-warts, mentioned elsewhere, they are probably to be regarded as sac-shaped prolongations of the outer layer.

Good evidence has not yet been brought forward of the existence of a nervous system in *Actinia*. A muscular apparatus is, however, well developed, and has been described in detail by M. Hollard. In the inner layer of the body-wall are two sets of flattened muscular fibres; a superficial circular, and a deeper longitudinal. Each mesentery has four muscles, two for each of its faces. The stomach wall is also provided with its own muscular fibres, these being so arranged in the vicinity of the inferior aperture as to permit the latter to be closed at pleasure. The existence of this sphincter is denied by some observers. A similar arrangement has been noticed in the delicate muscles which surround the tips of the tentacula.

Though, histologically, the several structures of *Actinia* admit of being resolved into two foundation membranes, an ectoderm and an endoderm, yet each of these, more especially the former, manifests a tendency to differentiate into other secondary layers, so that several apparently distinct tissues are recognizable in the body of the adult animal. This is well seen in the column wall, the principal thickness of which is composed of the two sets of muscular fibres mentioned above. That portion of the ectoderm which serves as an external investment to this muscular wall appears to consist, in some Actiniæ at least, of two separable, transparent membranes; an outer, or epithelial, forming the general surface of the body, and an inner or dermal layer in immediate contact with the muscular substance. The dermal membrane is almost wholly made up of a structureless periplast containing very few endoplasts; in the epithelium, however, endoplasts are more abundant. Between these two membranes thread-cells are sometimes found embedded in such numbers as almost to form a true layer, while close beneath the epithelium occur masses of the pigment granules, to which the varied, and often gorgeous, colours of these animals would seem to be due. The endodermal lining of the muscular wall is, in like manner, composed of two membranes, the one superficial, the other in direct contact with the deeper longitudinal fibres. Both ectoderm and endoderm have their free surfaces more or less abundantly ciliated. The structure of the digestive sac does not differ, in any essential respect, from that of the column, than which it is much thinner and more delicate, its endoderm being richly furnished with cilia. Ordinary pigment granules are here absent, but in their stead occurs, within the upper portion of the stomach wall, a thin layer of a red or yellowish brown tint, to which some writers have ascribed the function of a liver. The mesenteries are to be regarded as processes of the column wall. The thin layers of endoderm which invest the two sides of each mesentery are produced beyond its free edge to form the sac-like covering, within which the reproductive elements are lodged. Having enclosed these, the two layers are brought into mutual contact, a narrow band being thus produced, to which the cord-like craspedum is attached.

The flower-like appearance of the fully expanded *Actinia* is sufficiently familiar to every sea-side observer. While the animal is in this condition any passing object likely to serve as food is firmly grasped by one or more of the tentacula, which, aided by the muscular contractions of the body wall, soon force it into the interior of the digestive sac. The morsel thus swallowed is usually, after a time, rejected by the mouth; while the nutritive matters withdrawn from its substance by the action of the stomach secretions are transferred to the somatic cavity, within which, as in the *Hydrozoa*, the process of nutrition is completed. Of the voracity of the *Actinia* many amusing accounts have been made known. It may, nevertheless, be kept in captivity for several months, if supplied with water containing minute particles of organic matter.

Contact of food, mechanical irritation, or the withdrawal of the sea-water within which it dwells, causes the *Actinia* to assume its contracted condition. In this state it appears as a somewhat conical, inert mass, often much flattened, the mouth and tentacles being more or less completely concealed by the folded margin of the disc. In the act of expanding, this is gradually rolled backwards, displaying the tentacles, which, as the margin continues to unfold itself, are soon distended to their full extent by the pressure of the fluid contained in the somatic cavity.

The *Actinia* has the power of effecting considerable alterations in the general form of its body by the alternate contraction and expansion of the muscular fibres mentioned above. It can also, like the *Hydra*, shift its position at pleasure, though some species, under ordinary circumstances, attach themselves so firmly as not to be removed without laceration of the base.

2. **General Morphology.**—In no essential respect does any Actinozoön depart from the typical structure above described, nor do the members of the present group present such varied modifications of a common plan as have been shown to appear in the *Hydrozoa*. In all *Actinozoa* the digestive apparatus, though communicating freely with the somatic cavity, is furnished with a wall of its own, between which and the outer boundary of the body the generative elements are produced. By these characters they may readily be distinguished from the *Hydrozoa*, with which, in the more minute details of their structure, they closely agree; the body of an Actinozoön, like that of

a Hydrozoön, wholly consisting of ectoderm and endoderm.

The entire class is divided into four orders. In the first of these, *Zoantharia*, represented by *Actinia* and its immediate allies, the number of mesenteries, tentacles, and other parts in connection therewith is, in general, some multiple of five or six. In the three remaining orders some multiple of the number four appears to prevail. Thus in the *Alcyonaria* there are eight somatic chambers, eight mesenteries, and eight tentacles, not simple, as in *Actinia*, but furnished with pinnate margins (*fig.* 26, *a* and *b*). The members of the third order, *Rugosa*, known only through its fossil representatives, seem to have possessed a structure in some respects intermediate between that of the two preceding sections. Lastly, the *Ctenophora* are free-swimming, gelatinous animals, in physiognomy widely different from the other *Actinozoa*, though evidently akin to these in their leading anatomical features. They are the most highly organized of Cœlenterate animals, a common representative of the order, *Pleurobrachia*, possessing a complex nutritive apparatus, together with well-defined organs for prehension, reproduction, locomotion, and, in addition, unmistakeable indications of a nervous system (*figs.* 27 and 39).

Like the members of the preceding class, many of the *Actinozoa* multiply freely by gemmation, complex plant-like individuals being thus formed which consist of numerous zoöids united by a cœnosarc (*figs.* 34, *d* and 35). In such instances, each nutritive zoöid, or that portion of the organism which answers to the polypite of a Hydrozoön, is distinguished by the name of 'polype.' When gemmation does not occur, as in several species of *Actinia*, the name polype is often employed to denote the entire animal.

Though the soft parts of the *Actinozoa* have only of late years received proper attention from zoölogists, yet the hard structures to which these animals give rise have, under the general name of "Corals," been objects of interest from a very remote period. The outward aspect of Corals, as preserved in our museums, is familiar to most persons. Their true nature, in relation to the living organisms by which they are produced, is known only to the student of the *Actinozoa*.

The limits in size presented by the several forms of *Actinozoa* are not very readily defined. The polypes of this group are usually much larger than the polypites of the *Hydrozoa*, and, in a few cases, attain a diameter of even eighteen inches. The gigantic dimensions of some of the coral structures, produced by a combined process of growth and gemmation, are well known. Though the separate polypes of such a mass may, in certain instances, be little larger than pin's heads, yet, very often, they are half-an-inch in diameter, and not unfrequently, their size is much more considerable. It can scarcely be said that any *Actinozoa* are of microscopic dimensions. All the *Ctenophora* are conspicuous animals; *Pleurobrachia*, already alluded to, one of the smaller members of the group, being often about the size of an ordinary marble.

The various structures of the *Actinozoa* may be described under the general heads of

  a. Organs of Nutrition,
  b. Prehensile apparatus,

c. Tegumentary organs,
  d. Corallum or Skeleton,
  e. Muscular system and organs of Locomotion,
  f. Nervous system and organs of Sense,
  g. Reproductive organs.

3. **Organs of Nutrition.**—The whole interior of the polypes and, in the budding species, that of the cœnosarc by which they are connected, constitutes the nutritive apparatus of the *Actinozoa*.

In most *Zoantharia* the structure and functions of the polypes are best illustrated by reference to the account of *Actinia*, above given. But in some members of this order the digestive sac is relatively much shorter than the somatic cavity, being, according to Dana, little more than ·2 of its length in the genus *Palythoa*.

So, likewise, among the *Alcyonaria*, the somatic cavity of each polype usually appears as a long, somewhat slender, tube, in the upper portion of which the comparatively short stomach is, as it were, suspended; the proximal, or post-stomachal, region of the body cavity becoming gradually much narrowed. In the *Gorgonidæ*, however, the somatic cavity is shorter and slightly dilated towards its basal extremity.

There are two apertures to the digestive cavity of every Actinozoön; first, the mouth, and secondly, the proximal or inferior outlet, which opens freely into the somatic cavity.

In many, though not all, *Alcyonaria*, the somatic cavities of the separate polypes which make up the compound mass are prolonged into canals, freely communicating with one another, inosculating, and forming a sort of aquiferous system, within which the nutritive products circulate. In *Meandrina* and certain other *Zoantharia*, the general cavities of the polypes open by wide apertures into each other; but in very many forms of coralligenous *Actinozoa* it is erroneous to suppose that any connection, available for nutrient purposes, is maintained between the different polypes of the same compound stock.

Although, in a large number of *Actinozoa*, the somatic cavity has no communication with the exterior, save through the digestive sac or the free ends of the tentacles, when these are perforate, yet, among other members of the class, the existence of apertures in the body-wall seems to have been satisfactorily ascertained. Mention has already been made of the cinclides of *Actinia*, nor are these orifices wanting in several allied genera. In *Philomedusa*, according to Fritz Müller, twelve rows of such openings, which appear to the naked eye as minute pale dots, and capable of independent contraction, radiate from the posterior extremity of the animal. In the centre of this extremity, as also in *Peachia* and *Cerianthus*, exists a much larger aperture, or, rather, short canal, which the animal has the power of closing effectually whilst its somatic cavity remains distended with fluid. And Milne Edwards has shown that in *Corallium* the cœnosarcal canals communicate directly with the surrounding medium by means of numerous perforations in their walls. Lastly, to the category of structures now under consideration must be referred the 'apical pores' of the *Ctenophora*, whose nutrient system presents peculiar features which render it necessary that some account of *Pleurobrachia* (= *Cydippe*), as a typical example of the group, should in this place be given (*fig.* 39, *e*).

The body, or 'actinosome,' of *Pleurobrachia* is sub-spherical or melon-shaped, colourless, gelatinous, and perfectly transparent, but displaying, in sunlight, tints of a beautiful iridescence.

Two poles, an oral and an apical, mark the opposite extremities of the axis of the animal. The slightly protuberant mouth appears, when closed, as an elliptical fissure, presenting two flattened sides and two opposing edges.

Eight meridional bands, or 'ctenophores,' bearing the comb-like fringes, or characteristic organs of locomotion, traverse, at definite intervals, the interpolar region, which they divide into an equal number of lune-like lobes, termed the 'actinomeres.' But this division of the body does not extend into the immediate vicinity of the poles, before reaching which the ctenophores gradually diminish in diameter, each terminating in a point. Around the apical pole, in particular, may be noticed a somewhat oblong, depressed, area, distinctly circumscribed by the adjacent converging actinomeres.

The eight actinomeres are by no means equal in size, and, to understand their relations aright, it seems desirable to distinguish three principal kinds of these parts as the anteroposterior, the lateral, and the accessory actinomeres.

The two antero-posterior actinomeres, wider than their fellows, are opposite to each other and the edges of the elliptical mouth. At right angles to these, but in like manner opposite one another, lie the two lateral actinomeres, which, therefore, face the flattened sides of the mouth. The accessory actinomeres, slightly narrower than either of the preceding, serve to occupy the four interspaces which occur between the lateral and antero-posterior pairs.

The lateral actinomeres are further distinguished by the presence in each of a large sac, which opens obliquely, outwards and downwards, about midway between the equatorial region and the apical pole of the body. From this sac the animal has the power of protruding at pleasure a long, highly contractile, beautifully fringed, tentacle.

Immediately within the apical pole is situate a peculiar body, supposed to be an organ of sense, which is best termed the 'ctenocyst.' Upon this rests a nervous mass from which issue small filaments. The structure of these parts, as also of the prehensile, locomotive, and reproductive apparatuses are described in their appropriate paragraphs. At present let us chiefly notice, in connection with the form of the body, the arrangement of its somewhat complex nutrient system. (*fig.* 27.)

This system may be said to commence in the stomach, or digestive sac, a cavity having the general form of an elliptic cylinder, and extending from the mouth through the longitudinal axis of the body, for about ·6 of its entire length. Slightly contracting below, the digestive sac is seen to open into a much wider and shorter cavity, also axial in its direction, known as the 'funnel,' which gradually diminishes in diameter as it approaches the apical pole of the body, to terminate just above the ctenocyst and nervous mass. From the funnel three pairs of canals are given off.

ACTINOZOA. 145

Two of these, the 'apical canals,' very short and narrow, run directly downwards and outwards on either side of the ctenocyst, and soon open externally as the 'apical pores,' situate immediately beyond the margin of the 'apical area.' Some writers describe the apical canals as lateral, others as antero-posterior in their direction. Both opinions are partially correct, the apical pores

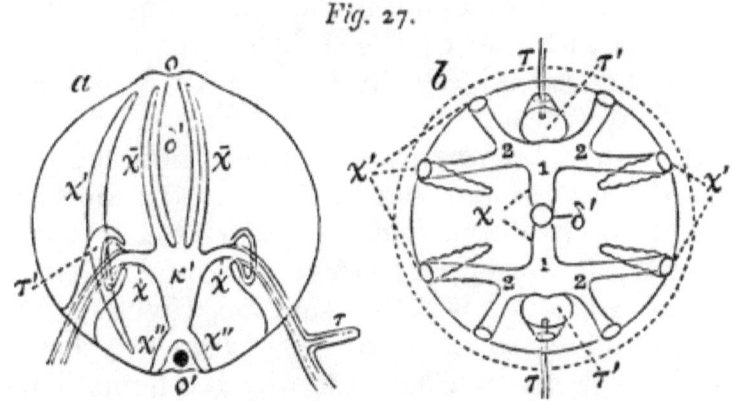

Fig. 27.

Morphology of PLEUROBRACHIA:— a, diagrammatic longitudinal section of *Pleurobrachia*; b, the same in transverse section; o, mouth; δ', digestive cavity; κ', funnel; o', ganglion and ctenocyst; τ', sac of the tentacle; τ, tentacle, with one of its branches; χ, radial canal; χ', one of the ctenophoral canals; χ'', apical canal; $\bar{\chi}$, paragastric canal. The numerals 1 and 2 indicate the first and second bifurcations of the radial canals.

being obliquely opposite one another, though placed on different sides of the body. Two other canals, the 'paragastric canals,' assume an upward course, parallel to and not far from the flattened sides of the digestive sac, but terminate cæcally before quite reaching its oral extremity. A third pair of canals, much wider and shorter than those just mentioned, radiate from the funnel in a

L

horizontal or slightly oblique direction, proceeding towards the bases of the pits in which the tentacles are lodged. Before gaining these, however, each 'radial canal' divides into two branches, the secondary radial canals; each of these again into two others, and, thus, eight tertiary radial canals are formed, which run towards the equatorial region of the body, where they open at right angles into an equal number of longitudinal vessels, the 'ctenophoral canals,' whose course coincides with that of the eight locomotive bands. These canals end cæcally both at their oral and apical extremities.

If, now, a comparison be made between this nutrient system and that of *Actinia*, the digestive sacs of the two organisms are clearly seen to correspond; in form, in relative size, and mode of communication with the somatic cavity. The funnel and apical canals of *Pleurobrachia*, though more distinctly marked out, are the homologues of those parts of the general cavity which in *Actinia* are central in position and underlie the free end of the digestive sac. So, also, the paragastric and radial canals may be likened to those lateral portions of the somatic cavity of *Actinia* which are not included between the mesenteries. Lastly, the ctenophoral canals of *Pleurobrachia* and the somatic chambers of *Actinia* appear to be truly homologous, the chief difference between the two forms being that while in the latter the body chambers are wide and separated by very thin partitions, they are in *Pleurobrachia* reduced to the condition of tubes; the mesenteries which intervene becoming very thick and gelatinous, so as to constitute, indeed, the principal bulk of the body. In both, the nutrient system is lined by a ciliated endoderm, the vibratory action of which serves to maintain a circulatory motion of the included fluid. The contractile tissues of the ectoderm may further assist such movements. And in *Pleurobrachia*, whose bilateral symmetry is more strongly marked than that of *Actinia*, the nutrient fluid, as Agassiz has shown, is at times alternately impelled between the right and left sides of the somatic cavity.

Very many curious modifications are presented by the canal system among the different genera of *Ctenophora*, to some of which reference will be made in the more particular account to be given of that order.

No manducatory apparatus exists in the *Actinozoa*. The oral margin may, however, be somewhat thicker and firmer than the surrounding parts, or otherwise become altered in appearance; and the cilia of the digestive sac may also differ from those which occur on other regions of the body.

As in *Actinia*, one part of the digestive cavity may undergo some amount of modification, coloured granular masses appearing in its walls which have been supposed to indicate a liver. Such coloured cells in the *Ctenophora* usually arrange themselves as vertical ridges which surround the innermost, or stomachal, division of the otherwise transparent digestive sac.

Milne Edwards has also shown the existence in *Cestum* of another structure whose function is probably secretive. Between certain of the ciliated bands and their corresponding ctenophoral canals; parallel to, and in close connection with the latter, runs a tube occupied by a number of granular bodies, and giving off, at right angles to itself, a series of short vertical branches, which open along the line of the locomotive fringes to communicate with the surrounding medium.

4. **Prehensile apparatus.** — The tentacles of the *Actinozoa*, like those of the *Hydrozoa*, appear usually, if not always, as hollow appendages, in immediate connection with the somatic cavity, their walls being richly provided with thread-cells and consisting throughout of two layers, an ectoderm and an endoderm.

Among the *Zoantharia*, the tentacles vary exceedingly in size and external form. Viewed from without, they are seen to arise, save in *Eumenides*, from the distal extremities of the polypes, between the mouth and the outer margin of the disc (*figs.* 33—35). Dissection shows them to be hollow processes in free communication with the somatic chambers, each of which is furnished with one or more of these appendages. Their most usual form is that of a slightly curved, more or less tapering, cone, as in many species of the genus *Actinia* itself. But from this typical aspect there are very many aberrant modifications.

Among the *Alcyonaria*, the tentacles are comparatively short, closely arranged in a single cycle of eight around the mouth of each polype, their margins being produced into a number of lateral pinnæ (*fig.* 26, *a*). These last, according to Dana, are perforate at their free ends, the extremity of the tentacle itself being cæcal; but this statement is denied by Milne Edwards and others, who more correctly view the pinnæ, in some genera at least, as destitute of distal orifices. The pinnæ are very contractile, so as to vary in form from mere lobes or tubercles to long filiform fringes. But little diversity is exhibited by the tentacles of this order. Except in the distinctive characters just mentioned, they agree essentially with those of *Actinia*.

The tentacles of the *Ctenophora* are best described in connection with the general survey of the characters of that order.

5. **Tegumentary Organs.**—In but few *Actinozoa* do the tentacles appear to be processes of the ectoderm only. This layer, as we have seen in *Actinia*, exhibits a tendency to differentiate into two diverse planes of growth, which, with Professor Huxley, we may designate the 'ecderon' and the 'enderon', respectively. Sometimes, however, this distinction is not observable. The ectoderm is usually ciliated, and in the *Ctenophora* becomes very thick and gelatinous, presenting a structure somewhat similar to that which occurs in the oceanic *Hydrozoa*. Gegenbaur describes the reticulating threads which traverse the periplastic mass as tubular in young *Ctenophora*, but, as growth advances, tending to become solid. Other minor histological modifications have been observed.

The general surface of the body, smooth in most *Ctenophora*, is in *Chiajea* and a few other genera diversified at intervals by the elevation of numerous simple papillæ. And, in some Sea-anemones, it exhibits a number of clear warts or vesicles, each of which, according to M. Hollard, possesses a muscular arrangement of its own, in connection with a sort of two-lipped mouth; so that a needle, or other small foreign body, introduced into the vesicle, is quickly and tenaciously secured. In their natural situations these creatures are often completely covered by fragments of shell, gravel, or sand, attached to their bodies by a peculiar viscid secretion, in the production of which these warts are, perhaps, concerned.

Or, the epidermic secretion may give rise to a distinct membranoid coat, protecting the integument of the animal, from which it is at times cast off by what may be termed a process of sloughing. Such a membrane in *Cerianthus* Mr. Gosse states to be " wholly composed" of altered cnidæ, which intertwine one with another to form a wide tube, investing the entire surface of the column. Here the connection of the tube is so loose that it can at any time be removed without much inconvenience to the animal, but, in other genera, a more adherent covering may be found. In *Adamsia* the base excretes a delicate, somewhat chitinous membrane, which, upon occasion, may continue its growth beyond the attached outline of its possessor, and even form an artificial extension of the peculiar surface which this genus is wont to choose for its abode.

The thread-cells of the *Zoantharia* have been studied with great care by Mr. Gosse, who distinguishes four principal kinds of these bodies by the titles of 'chambered,' 'spiral,' 'tangled' and 'globate cnidæ.' The chambered cnidæ (which are the most common) are of a long oval form, the ecthoræum, which varies greatly in length, presenting in all cases, the complex armature characteristic of these minute weapons; a number of delicate barbs, or 'pterygia', being attached to a thickened band, the 'strebla', twisted in a screw-like manner around the basal portion of the thread. The tangled cnidæ are relatively broader then the preceding, having a very long ecthoræum, loosely rolled up into a confused bundle. The spiral cnidæ present a much elongated, fusiform chamber, within which the thread lies coiled in a close regular spiral. Lastly, the so-called globate cnidæ have been seen to push out at each end a cylindrical protuberance, sometimes equal in length to the cnida itself, which does not contain any thread.

On the urticating organs of the *Alcyonaria* less attention has been bestowed. In general, they are of minute size and seem to resemble the tangled cnidæ of the *Zoantharia*. In *Sarcodictyon* they are aggregated on the tentacular pinnæ in minute rounded swellings, homologous with the palpocils of the fixed *Hydrozoa*.

The thread-cells of the *Ctenophora* present a peculiar structure. Each, in *Pleurobrachia*, according to Professor H. J. Clark, appears of a rounded or slightly napiform figure, and is covered externally by a single, dense, layer of very minute granules. From the summit of a broad conical projection on the inner surface of its otherwise uniformly thick, but rather delicate, wall, arises, in a very oblique direction, the simple thread, which, after making not more than seven or eight, equi-distant, spiral turns, set very far apart, terminates suddenly in what seems to be a free ending, precisely opposite its point of attachment. The thread is cylindrical, smooth, apparently solid, of firm consistence, and about eight or nine times the length of its envelope, from which it is set free by the gaping of the cell itself, around the thread's distal extremity.

On the whole it seems safe to say that among the *Actinozoa* the thread-cells exhibit a greater tendency to become collected in particular organs than has been shown to be the case with the *Hydrozoa*; though we by no means wish to forget the tentacles or nematophores of the latter. The mesenteric cords of the Sea-anemones strikingly illustrate this, and, in the *Ctenophora*, the urticating organs form a well marked layer on the outer surface of the tentacles and their lateral fringes. Parallel to, and agreeing in position with, these last, the two tentacles in *Hormiphora* are furnished, as Gegenbaur has proved, with a number of very peculiar, bright yellow, appendages, one between from about every ten to fifteen of the ordinary side filaments. Each of these bodies, which serve as special receptacles for the thread-cells, is hollow, of a flattened fusiform, or lancet-shaped, form, with a short stalk of attachment, above which it is prolonged laterally into several pairs of tubular processes, which gradually diminish in length, and finally vanish altogether, before reaching its free, simply tapering, extremity.

Pigment-masses, irregularly scattered in some *Actinozoa*, are in others combined so as to form more or less definite layers, which may readily be examined in the commoner species of Sea-anemones. In the substance of the body-wall and tentacles, outside the muscles of the mesenteries, or even in the digestive tube itself, such interrupted layers of colouring-matter have been observed.

The exquisite roseate tint of some *Ctenophora*

is due to the presence of pigment-streaks or less regular stellate masses, in various parts of the ectoderm.

6. **Corallum or Skeleton.**—Intimately connected with the tegumentary organs of these animals, under which head, indeed, it might without impropriety be described, is the so-called skeleton, or 'corallum', with which so many of them are furnished.

The term coral, or corallum, is properly restricted, in zoölogy, to the hard structures deposited in the tissues, or by the tissues, of the *Actinozoa*. Any form of this class which possesses such a framework is called a 'Coral'.

All *Actinozoa* are not coralligenous. The *Ctenophora* and several species of *Zoantharia* deposit no corallum. On the other hand, the order *Rugosa* is known only from the remains of extinct Corals.

Of coral structures there are two principal kinds, which must be carefully distinguished from one another. First, the 'sclerobasic' corallum, a true tegumentary excretion, formed by the conversion of successive growths from the outer surface of the ecderon. Secondly, the 'sclerodermic' corallum, which better merits the name of skeleton, deposited, as it is, within the tissues of the animal, and, in all probability, by the enderon.

The sclerobasic corallum is by Mr. Dana termed "foot secretion"; the sclerodermic, "tissue secretion".

Let us first notice the sclerobasic corallum, which is found only in certain budding composite

Actinozoa. Most frequently its texture is simply corneous, but in *Corallium* proper and a few other forms, it becomes calcareous by deposition; and in *Hyalonema* and *Hyalopathes*, if these be true *Actinozoa*, it is siliceous. In *Isis* and *Mopsea* it consists of alternately disposed calcareous and horny segments, thus, as it were, combining strength with a yielding pliancy. In *Isis* branches are developed from the calcareous, in *Mopsea* from the horny segments of the sclerobasis. *Melitæa* presents a like structure, save that, in it, the corneous segments are replaced by others which assume a porous and suberous aspect.

Section of a sclerobasis shows it to be, in some cases, solid or nearly so; in others, distinctly resolvable into concentric layers, which serve, also, to illustrate the manner in which it has been produced; while, more rarely, it is composed of an aggregation of separate fibres.

Two principal modifications of form distinguish the sclerobasis. In some *Actinozoa* it constitutes a free axis, virgate or pinnately divided, and varying much in composition and thickness. In others it is attached, simple or branched, and often singularly plant-like in physiognomy, as in those *Gorgonidæ* to which the name of Sea-shrubs has been applied.

The relations of such structures to the soft parts of the animal are, with little difficulty, discerned. The sclerobasic corallum is, in fact, *outside* the bases of the polypes and their connecting cœnosarc, which, at the same time, receive support from the hard axis which they serve to conceal. Thus the cœnosarc of these corals appears as a soft, fleshy covering, from which the several polypes arise, their somatic cavities freely communicating one with another.

Far different in its nature is the sclerodermic corallum, deposited, as above stated, within the bodies of polypes, which, in some cases, remain separate, but, in others, multiply by continuous gemmation. And, just as the whole body of an Actinozoön is made up either of one polype or of several united by a cœnosarc, so, too, may the fully developed sclerodermic corallum consist of a single 'corallite' or of several connected by a 'cœnenchyma'.

The parts of a typical corallite are these (*fig.* 28). First, an outer wall, or 'theca', somewhat cylindrical in form, terminating distally in a cup-like excavation, or 'calice', and having its central axis traversed by a 'columella'. The space between this and the theca is divided into 'loculi', or chambers, by a number of radiating vertical partitions, the 'septa'. These do not, in certain instances, quite reach the columella, but are broken up into upright pillars, or 'pali', arranged in one, two, or three circular rows, termed 'coronets'. All the preceding parts are best brought into view by transverse section. Longitudinal division of a corallite shows, frequently, the existence of imperfect transverse partitions, or 'dissepiments', which, growing from the sides of the septa, interfere, to a greater or less extent, with the perfect continuity of the loculi. Sometimes the septa have their "sides covered with styliform or echinulate processes, which, in general, meet so as to constitute numerous 'synapticulæ', or transverse props, extending across the loculi like the bars of a grate." In other cases, the dissepiments are replaced by the development of successive horizontal floors, or 'tabulæ', which do not grow from the septa, but

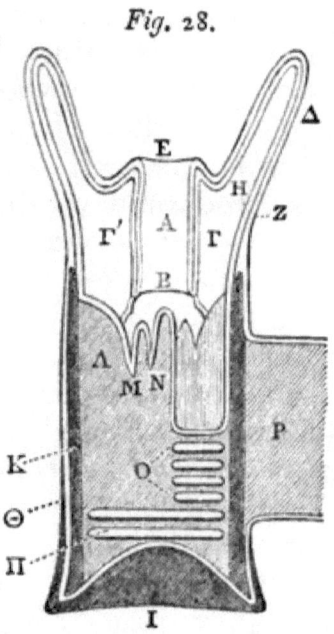

Fig. 28.

Morphology of ZOANTHARIA SCLERODERMATA:—A diagrammatic section of a living corallite. A, digestive cavity; B, its communication with the somatic cavity; Γ, intermesenteric chamber; Γ', mesentery; Δ, tentacle; E, mouth; Z, ectoderm; H, endoderm; Θ, epitheca; I, sclerobase; K, theca; Λ, septa; M, palulus; N, columella; O, dissepiments; Π, tabulæ; P, cœnenchyma. (The septa should be seen between the dissepiments, but are left out for distinctness' sake.)

extend, without interruption, across the entire space bounded by the theca. On the outer surface of the latter may occur 'costæ', or vertical lines, corresponding in position to the septa within; 'exothecæ', which arise from the sides of the costæ, thus representing the dissepiments; and a continuous layer, or 'epitheca', consisting of the coalesced, external, indications of tabulæ.

It needs scarcely to be stated that an organism producing such a structure as the foregoing must closely have resembled, in every essential respect, the *Actinia*, or typical polype, previously described. The relations of the septa and pali to the mesenteries, of the theca to the column wall, of the columella to that part of the enderon which forms the floor of the somatic cavity below the digestive sac, are, indeed, sufficiently obvious. The septa, too, like the mesenteries, are primary, secondary, and tertiary, according to their degree of approximation to the columella; the primary septa alone being in direct contact therewith. All these parts are, in the living animal, completely concealed by the soft integuments: the digestive sac, and much of the somatic cavity, especially its upper portion, performing, as in the soft-bodied species, their proper nutrient and reproductive functions (*fig.* 33).

In a similar manner is the cœnenchyma deposited within the cœnosarc (*fig.* 28). It may be united with the corallites at their bases only, thus forming a creeping expansion or stalk, or become connected with them throughout the greater portion of their height. There are even cases in which the corallites appear sunk amid a very abundant cœnenchyma, while, in others, the same structure is but sparingly developed. The relative distance of the corallites from one another is also subject to much variation.

But the typical structure of the corallite above described does not admit of being studied in any single species. Its nearest approach, as Milne Edwards has stated, is found in the genus *Acervularia*, which wants, however, synapticulæ and columella, the pali, also, being rudimentary. This genus is a member of the extinct order *Rugosa*, in which the sclerodermic corallum may, perhaps, be said to attain its most remarkable development. Both septa and tabulæ here occur in the same corallite, the former being always arranged in multiples of four.

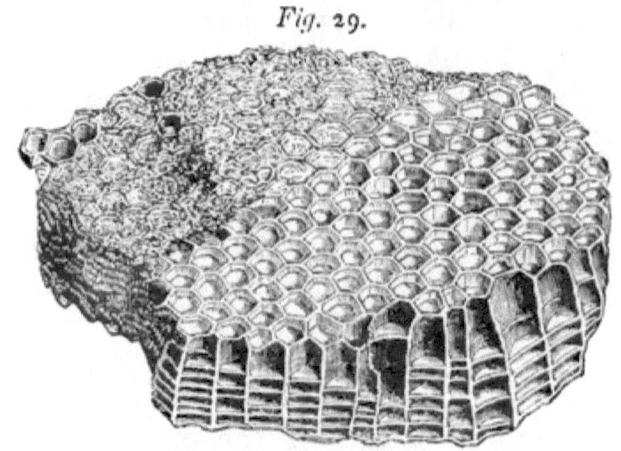

Fig. 29.

COLUMNARIA FRANKLINII.
Portion of corallum, of the natural size.

Among the sclerodermic *Zoantharia* tabulæ and septa are scarcely known to co-exist, a special section of this group, *Tabulata*, being distinguished by the nearly exclusive possession of the former (*fig.* 29). In two other large divisions, the *Aporosa* and *Perforata*, including several families, septa, in sets of five or six, normally occur, and in some are associated with dissepiments, more rarely with synapticulæ. In a fourth section,

Tubulosa, the septa are indicated by mere streaks (*fig.* 36, *c*). And in the *Tubiporidæ*, a family of *Alcyonaria*, septa are absent; each corallite being a simple tube, connected with the thecæ around it by horizontal plates, which represent the inner transverse floors of the *Tabulata* (*fig.*30).

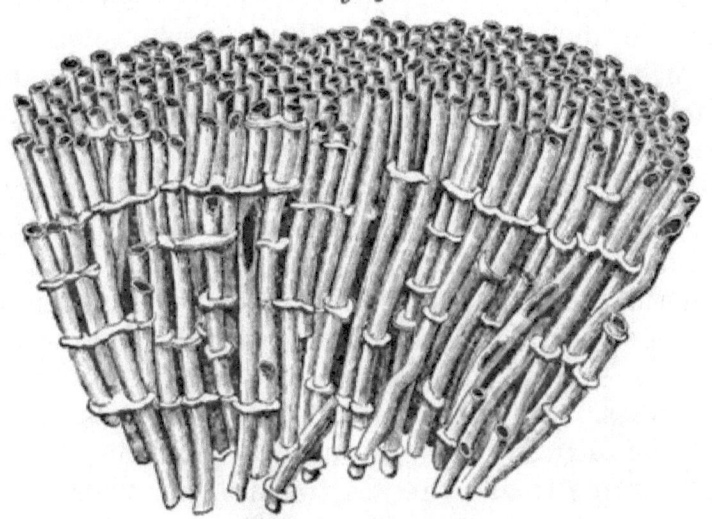

TUBIPORA MUSICA.
Fragment of corallum, of the natural size.

From the *Tubiporidæ* to other *Alcyonaria* in which the corallum, though sclerodermic, soon ceases to present traces of thecæ, a transition, not very abrupt, may be effected. Such intermediate stages, though not of much value to the systematic zoölogist, are of great interest in a morphological point of view, since they show well the manner in which the complete sclerodermic corallum has been formed; thus at once illustrating its minute structure and the several stages of its development. In *Telestho*, the corallum is made up of a number of branching tubes, which are not, as in all the preceding forms, perfectly calcareous. In *Cornularia* and its allies a corallum, never wholly tubular or of a firm calcareous consistence, has yet been detected; and in *Sarcodictyon* masses of spicules only can be observed. In some species of *Alcyonidæ* proper, the spicules attain a comparatively large size, and become aggregated into definite nodular masses. These 'dermosclerites', as Milne Edwards has shown, are of two principal kinds, the fusiform, and the irregular. The former are somewhat cartilaginous in consistence, and have their surface studded with slight asperities. The irregular nodules are stronger and more decidedly calcareous, presenting six faces, each, in general, furnished with a tubercular enlargement, which sometimes prolongs itself into a number of spines, bearing on their sides other secondary tubercles. By the coalescence of such masses and the deposition of more minute particles among their interstices, a thecal corallum, in other *Actinozoa*, at length comes to be formed. In *Alcyonium* itself the spicules, though numerous, are not of large size, and are most conspicuous in the column wall below the margin of the disc. Returning to the *Zoantharia* we find, in the genus *Zoanthus*, a spicular corallum still more feebly developed than that of *Alcyonium*. In many of the Sea-anemones no spicules have been observed, though traces of a corallum are not, even in these, absolutely lost. Finally among the *Ctenophora* we in vain search for the faintest indications of its existence.

From what has been said it were easy to infer that but little minute structure would be presented by the perfect sclerodermic corallum. Its decalcification, however, reveals delicate shreds of the periplastic substance by which it had been deposited, usually exhibiting an irregular reticulating arrangement. The 'sclerenchyma,' or coral tissue, presents every gradation between this nearly solid condition and the spicular stage permanently exemplified in *Alcyonium*. Thus, in the *Aporosa*, it is firm and compact; in the *Perforata*, porous and granular, or even spongy and reticulate.

In the accompanying table the chief modifications of the corallum, from an artificial point of view, are systematically exhibited.

It must not, however, be supposed that the presence of a sclerobasis renders the deposition of tissue secretions wholly impossible, for, among the *Gorgonidæ* it is certain that, in addition to the basal corallum, true sclerodermic spicules appear, within the substance of the investing mass. When such a *Gorgonia* is dried, and the soft parts washed away, a thin layer of calcareous spicules will be found gently adhering to the brown, horny sclerobasis below. M. Valenciennes has proposed to distinguish five kinds of these spicules, or 'sclerites,' by the names of capitate, fusiform, massive, stellate, and squamous, respectively.

KEY TO MODIFICATIONS OF CORALLUM.

Corallum wholly sclerodermic.
   Corallum thecal, calcareous.
      Tabulæ present.
         Septa in × of 4. . . . . Rugosa.
         Septa in × of 5 or 6, rudimentary or absent. . . . . . . Tabulata.
      Tabulæ absent.
         Septa well marked, in × of 5 or 6.
           Sclerenchyma porous. . . . Perforata.

Sclerenchyma imperforate. . APOROSA.
Septa indicated by mere streaks. Thecæ pear-shaped, in some connected by a basal, creeping cœnenchyma. . TUBULOSA.
Septa absent. Thecæ crowded, cylindrical, united at various heights by distinct, horizontal epithecæ. . TUBIPORIDÆ.
Corallum spicular or, if thecal, corneous or sub-calcareous.
Spicules numerous, in some replaced, either wholly or in part, by an imperfect, tubular corallum. . . . ALCYONIDÆ.
Spicules scanty, or replaced by particles of sand. . . . . . . ZOANTHIDÆ.
Corallum sclerobasic.
Sclerobasis spinulous or smooth. . . Z. SCLEROBASICA.
Sclerobasis sulcate.
Sclerobasis attached proximally. . GORGONIDÆ.
Sclerobasis free. . . . . PENNATULIDÆ.

7. **Muscular System and Organs of Locomotion.**—Reference has already been made to the muscular system of *Actinia*.

A like apparatus, presenting, however, some differences of detail, appears to become differentiated from the general periplastic substance in most other *Zoantharia* and *Alcyonaria*. But the power of altering the position of the body by the slow alternate contractions of a normally attached base is possessed only by those *Zoantharia* to which the name of Sea-anemones is usually applied. Their non-adherent allies, such as *Edwardsia* and *Cerianthus*, have a highly contractile columnwall, capable of greatly varying its length, and of executing movements, for the most part, of a feeble worm-like character. The *Alcyonidæ* and

Gorgonidæ are permanently fixed, as are also many of the higher coralligenous *Actinozoa*, especially those which multiply by continuous gemmation. Others, however, and these chiefly the simpler forms, are free, but, like the unattached *Pennatulidæ*, not truly locomotive. Yet in the greater number of the *Actinozoa* each polype, though fixed, is contractile to some extent, shrinking down under irritation, and again unfolding itself at pleasure, while, among the *Alcyonaria*, with a few exceptions, it is also retractile into the fleshy substance of the cœnosarc. Even this, too, has its own share of contractility, most evident in those species which possess an elastic sclerobasis. Thus, on the South American coast, Mr. Darwin observed a Sea-pen which, on being touched, forcibly drew back into the sand some inches of its compound, polype-covered, mass.

All the *Ctenophora* are free-swimming animals, but doubt yet hangs over the nature of certain exceptional *Zoantharia*, reputed to be of similar habit. The apical extremity of the genus *Minyas* and its allies is represented by Lesueur and Lesson as dilated into a large air-sac, excavated beneath the floor of the somatic cavity, and furnished below with an opening into the surrounding medium. By means of this sac the creature is said to float without effort, its oral disc being turned downwards; but further observations on its structure are much wanting. Again, the *Arachnactis albida* of Sars, possesses, according to Professor E. Forbes, not merely the power of swimming like a Medusid, but "it can convert its posterior extremity into a suctorial disc, and fix itself to bodies in the manner of an *Actinia*." But the aspect of the tentacles in this organism strongly suggests the possibility of its being an immature form, nor is the suspicion weakened by the discovery of Haime, that the young of *Cerianthus*, while resembling *Arachnactis* in physiognomy, enjoys a similar oceanic mode of life.

The muscular fibres of the *Actinozoa* are interesting to the histologist, as wanting, among many forms, those distinct transverse striæ, which, elsewhere, they so frequently present. Such striæ are not, however, always absent. In the body-substance of this class we have, in truth, obvious transitions from a simple contractile periplast to muscular fibres, which in no essential respect differ from those of various invertebrate animals. In the typical *Ctenophora*, the contractile tissues appear to be disposed in two principal sets; a transverse or circular, and a longitudinal.

Some *Zoantharia* employ their tentacles as aids to locomotion, though neither in these nor in the *Alcyonaria* can it rightly be said that special motile organs exist.

Of this nature, however, are the 'ctenophores,' or ciliated bands, which constitute so obvious a feature in the physiognomy of the *Ctenophora*. The normal number of these bands would seem to be eight, though in *Cestum*, and one or two other forms, their typical structure and arrangement is somewhat modified. Each ctenophore is of a much elongated ovate form, widest at the equatorial region of the body, and tapering gradually to end in a point at some distance from the oral and apical poles; slight differences in degree of approximation to these parts, and such-like minor characters, distinguishing the ctenophores of the several genera. The surface of the ctenophore is transversely elevated at intervals throughout the greater portion of its length into a number of successive ridges, to each of which a row of strong cilia is attached in such a manner as to form a paddle-like plate, or comb, the free extremities of the cilia remaining separate. The cilia are not all of equal length, those of the middle portion of the comb usually having the advantage in this respect, while the cilia on either side symmetrically correspond; their degree of elongation varying so as to impart to the edge of the entire comb a gently curved outline, when seen at rest. This is, indeed, seldom the case during the life of the animal, throughout which the combs manifest an astonishing amount both of simultaneous and successive activity. Nay, even after death, detached portions of these creatures, bearing fragments of the ctenophores, exhibit for many hours no apparent diminution of their ordinary vibratile efficiency.

8. **Nervous System and Organs of Sense.**—In no *Actinozoa*, save the *Ctenophora*, has good evidence of the presence of a nervous system or organs of sense yet been obtained. Nor should this appear surprising, for the sensitivity which, in more highly differentiated organisms, has its course restricted to definite tracks, is here diffused, in a less appreciable manner, through the more general and comparatively ill-developed tissues of the body. The white or blue marginal sacs of some Actiniæ, and the body-warts in allied species, have, it is true, been regarded as sensitive in function, and the former have even been dignified

by the title of rudimentary eyes. The radiating system of ganglia and nerve-fibres which Spix described as existing within the base of *Actinia* has not come under the notice of other observers.

But in the *Ctenophora* occurs a well-marked sense-organ, the 'ctenocyst,' upon whose precise function, whether oculiform or auditory, naturalists are far from being agreed. Such differences of opinion are in truth based on the prejudices which most anatomists acquire from a too exclusive attention to the structural peculiarities which the higher animals present. The ctenocyst, in all probability, neither sees nor hears, but would seem to be the localised recipient of those obscure general impressions to which its lowly-organised possessor is capable of responding.

The ctenocyst occupies a central position amid the soft substance of the ectoderm, immediately within the apical pole of the body. In form it is ovate or spherical, smooth externally, but, in some cases, invested with an adventitious layer of pigment granules. Its wall appears to be very firm and elastic, so as quickly to recover its proper figure, should this be changed in accordance with the ordinary contractions of the body. Within, the ctenocyst is hollow, and apparently distended with a fluid. In the midst of this fluid lie a number of rounded or polyhedral concretions, semitransparent, colourless or somewhat tinted, occasionally coalesced into a single mass, and composed, probably, of carbonate of lime. Each granule is little more than ·0003 of an inch in diameter. The concretions appear subject to a peculiar vibratory movement, but some observers have disputed the fact of its occurrence.

The nervous system of the *Ctenophora* consists either of a single ganglion or of a pair of ganglia closely approximated, giving origin to a number of delicate nerve-like cords. The ganglion lies deeply seated within the pyramidal mass of ectoderm included between the apical canals, towards the narrow extremity of which its apex is directed, while its base rests upon the surface of the ctenocyst. In form it is sub-pyriform or bluntly conical: anatomically, it seems resolvable into a thin transparent wall, enclosing granular contents; in colour, it is most frequently pale yellow. From this central mass issue two principal series of nervous cords, one of which arches inwards towards the walls of the digestive cavity, and, in some cases, separates into four sets or bundles to supply the principal regions of the body. The nerves of the second series, usually eight in number, are distributed along the rows of swimming combs so as to lie between the latter and their corresponding canals. These cords appear dilated at intervals into numerous minute ganglionic enlargements, giving off secondary filaments, one for each of the ciliated plates; an arrangement, which, if corroborated by subsequent investigations, would go far to throw some light upon the singular and quasi-independent movements which these combs perform in the living animal. There is still, however, much diversity of opinion as to the true interpretation of the parts just described. Kölliker, while recognising in *Chiajea* the presence of delicate cords extending from comb to comb, expresses himself, nevertheless, as very doubtful of the existence of a nervous system in any of the *Ctenophora* which he had himself investigated. Agassiz is equally sceptical. On the other hand, the careful observations of Will, Milne Edwards and, more recently, of Gegenbaur, point to an opposite conclusion. Somewhat similar are the views of Frey and Leuckart. All the preceding writers are unanimous in rejecting the prior account of the nervous system of *Pleurobrachia* given by Grant, who describes a double nervous ring surrounding the mouth, in the course of which he thought he could detect eight ganglia, each giving off on either side two fibres and a fifth larger filament, traceable onwards beyond the middle of the body. Yet this description, when carefully considered, is less irreconcileable with the views expressed on the same subject by other observers than seems to be usually supposed.

Certain curious appendages, which possess, perhaps, a tactile function, have been observed by R. Wagener in two genera of *Ctenophora*, *Beroe* and *Pleurobrachia*. These organs appear as long hair-like threads, which arise from either side of the ctenophores along their whole length, and form around each pole of the body a sort of wreath, composed of several concentric rings. The threads have not been seen to exhibit any independent movements. Each swells once or twice in the course of its length into a smooth or angular, rounded or flattened, expansion, the entire surface of which is abundantly beset with minute stalked knobs. In some threads smaller swellings, wanting the capitate stalks, take the place of those just noticed. Occasionally the threads branch and, instead of ending in points, terminate in dilatations of a like nature to those which interrupt their filiform axes.

9. **Reproductive Organs.** — The reproductive organs, in most *Alcyonaria* and *Zoantharia*, agree, both as to position and structure, with the same parts in *Actinia*; each spermarium or ovarium consisting of a prolongation of the peritoneal membrane which clothes the sides of the mesenteries, and forms along their free edges a double band, within which true generative elements are produced (*fig.* 26, *c*). Each band usually contains only ova or spermatozoa, but ovaria and spermaria may occur either in the same or in different polypes. *Cerianthus* and some other forms are monœcious, but more frequently, as in the majority of Sea-anemones, the sexes appear to be distinct. Not so, however, the *Ctenophora*, in which group bisexuality may certainly be said to prevail. An ovarium and spermarium occur as thickened folds along the opposite sides of each ctenophoral canal, beneath its endodermal lining. But the same mesentery gives rise on both of its free sides to only one kind of generative element. Here, as in other *Actinozoa*, the male and female organs differ in their contents alone.

The ova of the *Actinozoa* are, in general, of a rounded form, smooth or dilated, and often brilliantly coloured. Their structure is typical, presenting the parts common to ova in general. The spermatozoa are caudate, with a broadly conical or even heart-shaped body, to the apex of which the tail is usually attached.

Reproduction does not, as in so many *Hydrozoa*, devolve upon specially modified zoöids. Some *Actinozoa* have been known to become everted and die shortly after the maturation of their genital products. But in others, no such exhaustion seems to occur. Fertilisation is probably effected while yet the ova remain within the somatic cavity of the parent. Here, in many cases, the early stages of development also take place.

Section II.

Development of Actinozoa.

The life-history of the *Actinozoa* presents a series of phenomena by no means so diversified as those which have been shown to characterise the developmental cycle in most of the *Hydrozoa*. For here the embryo, by an easy and gradual succession of changes, tends finally to assume the condition of an organism similar to that which brought it forth.

In the evolution of the embryo the whole or a greater part of the fertilised ovum seems to be concerned. The product of the reproductive act usually soon appears as a ciliated body, while yolk-cleavage, division into layers, and formation, by liquefaction, of an internal nutrient cavity, take place in the ordinary manner.

Zoantharia and *Alcyonaria*.—Among the *Zoantharia* and *Alcyonaria* the further development of the primitive polype into which the embryo is resolved would seem to be, in most cases, as follows. The young animal, still retaining its cilia, assumes a somewhat oblong ovate form. A central depression then makes its appearance at one extremity, indicating the rudimentary mouth. The internal cavity enlarges. Meanwhile, the tegumentary system is gradually becoming distinct, thread-cells accumulating to form a superficial layer, beneath which pigment globules may also be observed. Soon the muscular substance begins to be differentiated, its longitudinal fibres and the first rudiments of the mesenteries being, at an early period, discoverable. Next, small protuberances arise round the mouth, each of which gradually elongates to form a tentacle. The initial number of tentacles in the embryonic polype always bears some relation to that observable among the adult forms of the group of which it is a member. Thus, in most *Zoantharia* either five or six tentacles first sprout forth, but this number is rapidly doubled, an increase in size of the older tentacles being simultaneously effected. But, in very young *Alcyonaria* eight tentacles appear, as in the mature polype.

Within the body-substance transverse contractile tissues may now, at length, be detected. Minor changes of external form also take place; the cilia disappear, or are replaced by others of smaller size; and the proximal extremity modifies itself in accordance with the habits of the adult animal.

But before the formation of its tentacles, the young polype undergoes that important structural change which distinguishes it from the rudimentary polypite of the *Hydrozoa*. A circular fold of the body-substance surrounds the oral extremity and grows inwards in such a manner as to produce the wide digestive sac, open above and below, and freely communicating with the somatic cavity, from which it, nevertheless, remains distinct. The histological composition of the wall of this sac differs, in no essential respect, from that of the outer boundary of the body.

The polype, while yet immature, presents a well-marked bilateral symmetry. The oral fissure is produced more in one direction than in another, its form being by no means, as some have wrongly stated, circular. At its opposite angles, gonidial grooves, in certain cases one only, arise. Two of the mesenteries in *Actinia*, as Haime has pointed out, are developed opposite to each other before the rest make their appearance, and these in direction correspond with the two mouth-angles. The mesenteries grow from above downwards and, in some long-bodied polypes, do not extend much farther than the level of the free end of the digestive sac, or, becoming narrowed and much convoluted, are finally lost in the proximal portion of the wall of the somatic cavity. In *Cerianthus* two of the mesenteries descend, far below the others, almost to the orifice at the base of the general cavity. The remaining mesenteries, much shorter than the preceding, gradually diminish in length till they reach two points at either side of the larger mesenteries and, like them, opposite to one another.

The rudimentary tentacles, also, afford proofs of the symmetry just noticed. In those young *Zoantharia* which possess five of these appendages, four, as Agassiz has stated, are arranged in pairs on either side of the mouth, while the fifth lies opposite one of the oral angles.

The subsequent development of the tentacles has been well illustrated by Haime from the case of the common Sea-anemone. The succession of these organs is effected from within outwards in a series of concentric circlets, each of which, save the second, includes twice the number of tentacles proper to its predecessor. Thus, the first circlet contains 6 tentacles, the second 6, the third 12, the fourth 24, the fifth 48, and the sixth 96. In *Antipathes* and some other polypes never more than the first six tentacles arise. Among certain other *Zoantharia*, one or more tentacles are occasionally aborted, and hence the somewhat puzzling numerical proportion of these organs noticed in slightly abnormal forms of this group.

Should the young polype give rise to a calcareous corallum, the early stages of its deposition consist chiefly in the formation of spicules which, at first small and detached, gradually increase in size and coalesce in a greater or less degree to form the various structures whose nature has elsewhere been explained.

Where a septal apparatus occurs, the development of its several elements follows the same definite law by which the number of the mesenteries and tentacles is determined.

The young Zoantharian has at first six septa, the Rugose Coral four, equidistant from one another, and separating similar loculi. In a few genera only can a smaller number of initial septa be detected. But among the great majority of *Zoantharia*, the typical grouping of the septa is hexameral; among the *Rugosa*, tetrameral.

In some *Zoantharia* the number of the septa never exceeds six, but, more frequently, new septa appear midway between those first formed, at the lower portion of the theca. These gradually grow inwards, at the same time increasing in height. Thus the primary loculi become divided into secondary chambers; these, by the formation of other intermediate septa, into tertiary chambers, and so on till the development of the corallite has been accomplished. The primary loculi alone are complete, for the septa limiting the other chambers neither extend so high, nor so closely approach the columella, as do those which are afterwards formed; the size of all succeeding series of septa being in direct ratio to the order of their development. So that the latter may be almost as clearly pronounced in an adult corallite as in a collection of specimens of different ages belonging to the same species. In like manner those stages of the septal apparatus which are transitional among the more complex corallites are well illustrated by an appeal to the various permanent conditions of the same system among less developed representatives of the group. Thus in some species of *Stylophora* we can count but six septa, and an equal number of primary chambers. In a much larger number of *Zoantharia* twelve septa may be observed, of which six are primary and six secondary. In others, there are twenty-four septa; six primary, six secondary, and twelve tertiary. The septal formula in all these types admits, therefore, of being, respectively, stated thus:

$$\text{I.} \quad 6 \qquad\qquad = \text{S. } 6.$$
$$\text{II.} \quad 6+6 \qquad = \text{S. } 12.$$
$$\text{III.} \quad 6+6+12 = \text{S. } 24.$$

in S. 12 *one*, and in S. 24 *three* additional septa being developed between each of the primary pairs. And, since the normal number of primary septa is 6 among the *Zoantharia*, it might be expected that for all corallites of this order having a higher formula than S. 24, the number of new septa produced within each of the first formed chambers would be represented by the successive terms of the series 7, 15, 31, 63, &c., in which each number is double plus unity of its antecedent. Practically, however, we find that, as soon as the formula S. 24 has been reached, only two septa arise at the same time in each primary chamber, or, in other words, the corallite developes simultaneously not more than twelve additional septa. So that the simple expression $n \times 6 + 6$ at once determines the normal septal formula for all Zoantharian corallites having twenty-four or more septa. Here n is $=$ the number of septa between each primary pair, and corresponds to the successive terms of the arithmetical progression 5, 7, 9, 11, 13, &c.

Since, therefore, the number of septa in process of formation is often less than the number of loculi, it becomes necessary to determine those chambers in which the new septa first appear, and the precise order of succession which they observe. But first let us explain the few technical terms by which the facts to be announced have been rendered susceptible of definition.

All septa which commence their growth simultaneously are said to be of the same *order*, while those which divide chambers of equal size belong to the same *cycle*. It is evident that septa of the first three cycles must correspond to those of the first, second, and third orders, respectively. But the fourth cycle includes septa of the fourth and fifth orders; the fifth, of the sixth, seventh, eighth, and ninth; and the sixth of the 10th, 11th, 12th,

13th, 14th, 15th, 16th, and 17th orders. In but few corallites does a seventh cycle occur.

The lines in the subjoined diagram are supposed to represent the septa of part of a corallite having the formula S. 48, each septum being indicated by the numeral proper to its order.

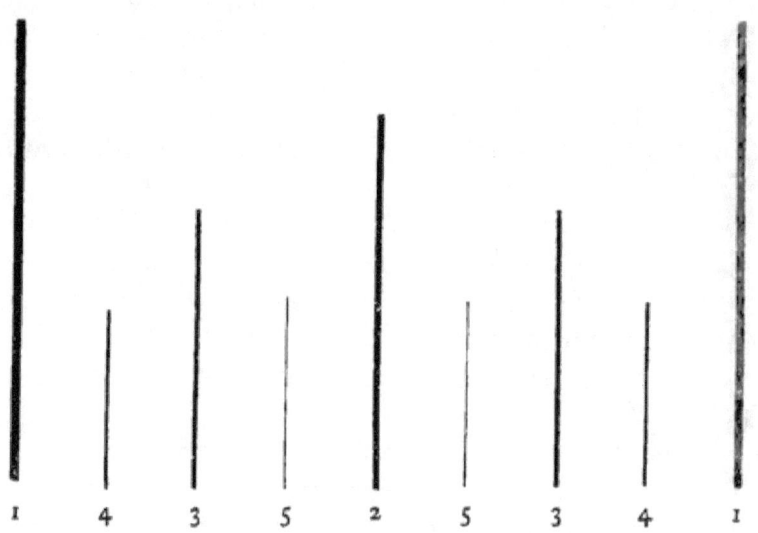

The same numerals also enable us readily to point out any one of the loculi included between the primary septa. In the present case the chambers, from left to right, would be denoted as 1 + 4, 4 + 3, and so on, respectively. Again, chambers of equal size are said to be *similar*, and those represented by corresponding numerals are of the same *expression*. Thus, in the diagram, the chambers 1 + 4 and 4 + 1 have the same expression. So, also, just before the development of septa of the fifth order, the chambers then separated from one another by the septa marked 3 could not have been denominated similar.

We are now in a position to understand the five rules of septal development laid down by Milne Edwards and Haime, to whom science is indebted for most of what we yet know on the subject under consideration.

Rule 1. The formation of new septa takes place *simultaneously* in all loculi having the same expression.

Rule 2. The formation of septa takes place *successively* in loculi having a different expression.

Rule 3. The order of succession of the septa is determined in the first place by the age of the cycle to which they belong, and those of a new cycle do not commence to be formed till the development of the preceding cycle is complete.

Rule 4. Among loculi of the same cycle having different expressions precedence in the formation of new septa is determined by the inferiority of the sum of the two terms of this expression.

Thus, if septa of the 6th order were to be developed in the corallite indicated by the above diagram, we should expect them to appear in the chambers $1+4$ and $4+1$.

Rule 5. Among loculi of the same cycle having different expressions, but which yield the same sum by the addition of the two terms of each expression, the order of appearance of the septa is determined by the relations which exist between the lowest terms of these expressions, the new septa being formed first where the lowest term occurs.

Or, recurring to our diagram, we might look for the appearance of new septa in the chambers $5+2$ and $2+5$ sooner than in $4+3$ and $3+4$.

As might be expected, from abortion and other causes, variations from the above arrangements now and then occur, the careful investigation of which is far from exciting that attention which it so well deserves.

It is to be remarked that some coralligenous polypes, when in confinement, appear to attain a considerable bulk before any traces of their skeleton can be observed.

Except in its minute size, and the comparative paucity of its tentacula, the young polype, when first excluded, closely resembles its parent, through whose mouth it usually makes its entrance into the surrounding world. The student will experience little difficulty in obtaining Actiniæ containing young in several stages of development for detailed anatomical examination.

It is to be wished that the embryology of the composite *Alcyonaria* and *Zoantharia* were more efficiently worked out, since it is just possible that, apart from other modifications of the developmental process, a rudimentary cœnosarc may, under certain circumstances, be produced before the formation of a distinct polype.

CTENOPHORA. The development of the *Ctenophora*, while presenting some peculiar features, resembles that of other *Actinozoa* in the comparative rapidity with which its early stages progress, so that the product of the reproductive act, while yet of small size, attains a form and structure similar to the parent.

The young *Bolina*, according to Mc Crady, is not very unlike an adult *Pleurobrachia* in shape, but the oral extremity appears somewhat truncate, and smaller than the more rounded apical region. From around the latter radiate eight short ctenophores, each containing from five to seven combs. The mouth leads into a very large digestive sac, tapering rapidly as it proceeds inwards to meet the narrow funnel, from which are given off two broad lateral sinuses, opening by eight "short pointed projections" into the embryonic indications of the ctenophoral canals. The extremity of each horizontal sinus also connects itself with a wide excavation, or pit, at the side of the body, from which freely issues a short tentacle, furnished with three or four fringing threads. Apical canals, also, may be observed to the right and left of the large rudiment of the ctenocyst. At a later stage are developed the paragastric canals, while as yet the ctenophoral tubes remain short and incomplete. These last are seen to approach the apical region, before lengthening so as to come near the oral pole of the body. Their development would seem to be somewhat in advance of that of the ctenophores themselves. From the wide horizontal sinuses branching radial canals soon shape themselves out. The mouth gradually becomes depressed, while the large antero-posterior lobes of the adult animal are making their appearance. Meanwhile, the tentacular pits shift, as it were, rather closer to the oral extremity, until at length they arrange themselves in their final position on either side of the mouth. The tentacles alter very much in appearance as their development proceeds, becoming more numerous, but simple and less contractile, so as no longer to resemble those of *Pleurobrachia*. Short branches are given off from the extremities of the paragastric canals, and from these the two canals which run by the sides of the mouth are probably produced. Not until the large lobes have become very distinctly recognizable do the earlets, four small appendages in connection with the lateral ctenophores, render themselves visible.

The observations of Semper on *Chiajea multicornis*, another lobed representative of the same order, indicate a still more rapid evolution of the embryo, which before quitting its egg-covering has the outward form of the adult animal, the canal system and ctenophores being as yet rudimentary. As in *Bolina*, the digestive tube appears the first formed part of that system, and is, indeed, developed earlier than any other internal organ. At no period of its career is the young *Chiajea* provided with a uniform covering of cilia.

In *Beroe*, a genus destitute of tentacles, the funnel of the embryo is comparatively large, before the ctenophoral canals are fully developed. The circular oral vessel is formed from two lateral tubes, whose extremities anastomose with four of the ctenophoral canals, while as yet the four others have not approached more than half way the oral pole of the body, towards which, as in the preceding, they are gradually developed. We have here an interesting proof of the bilateral symmetry of the *Ctenophora*, for in the adult *Beroe*, as from its want of tentacles might have been expected, this symmetry is, on a hasty inspection, less obvious than in most other members of the order.

Like *Chiajea*, *Pleurobrachia* is developed within

an egg-covering, and with an equal degree of rapidity. After yolk-cleavage the embryo appears rudely cylindrical in form, a belt of cilia passing round the middle of its body (*fig.* 31). This soon

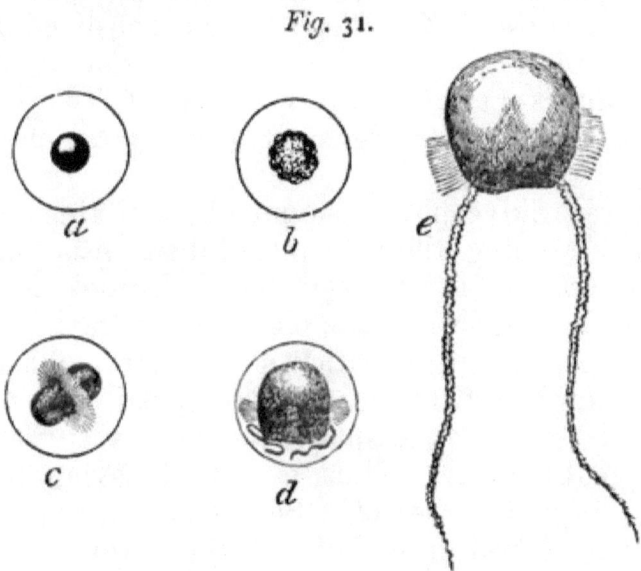

Fig. 31.

Development of PLEUROBRACHIA:—*a*, newly extruded ovum, showing the yolk with its external covering; *b*, the same, after segmentation; *c*, *d*, and *e*, successive stages of the embryo. (All magnified.)

breaks up into two lateral groups which eventually disappear altogether; the ctenophores, at first very broad and few in number, at an early period taking on the performance of their special function. The tentacles are at first destitute of lateral fringes, traces of which appear while the rest of the canal system is but imperfectly indicated. The form of the adult animal is, according to Agassiz, fully assumed before the young organism attains a length of ·04 of an inch.

Of the duration of life among the *Actinozoa*, as among *Cœlenterata* in general, we are still very ignorant. Some *Ctenophora* appear to last but a single season, yet this statement can by no means be regarded as true of the entire group. The *Alcyonaria* and *Zoantharia* seem long-lived and hardy animals, a specimen of the common Sea-anemone, kept in confinement for forty years, showing no visible signs of decrepitude or old age.

In reparative power the members of this class, notwithstanding their increased differentiation of tissue, appear fully to rival the *Hydrozoa*. Many experiments show how complete may be the healing of wounds and regeneration of lost parts among a large number of *Actinozoa*. Occasionally the process of reparation displays itself in a curiously abnormal manner. Thus, a section having been made below the disc of a Sea-anemone, tentacles were developed from both of the fresh surfaces thus exposed.

A common British Zoantharian, *Anthea cereus*, adapts itself well to the conditions of such experiments. Artificial fission of this species, if performed with care, does not always result in death of the parts so divided.

The same animal may also illustrate the mode in which spontaneous fission occurs among many other forms of *Actinozoa*. A longitudinal cleavage of the polype commences, in most cases, across the region of the disc, and thence proceeds downwards towards its proximal extremity.

Less frequently is fission effected by the separation of small portions from the attached base of the primitive organism, whose form and structure they subsequently, by gradual development, tend to assume.

Further observations are wanting on the occurrence of fission and gemmation among the *Ctenophora*. In none of these animals do we find colonies of zoöids resulting, as in the *Hydrozoa*, from a process of continuous budding.

But in other *Actinozoa* continuous gemmation abundantly takes place, and in this manner are formed those composite structures, consisting of numerous polypes, met with among so many genera of *Zoantharia, Alcyonaria* and *Rugosa*. In a few of these organisms, discontinuous gemmation may also be noticed.

Young polypes may be budded either (1) from the base of the primitive structure, or (2) from the sides of the polypes, or (3) from their oral discs. Since these surfaces are but parts of a common integument it might be anticipated that intermediate positions of buds would now and then occur. Nevertheless, it has been found convenient to distinguish three principal modes in which gemmation of polypes may be effected, as the *basal*, the *parietal*, and the *calicular*, respectively.

In basal gemmation the polype sends forth a rudimentary coenosarc, from which, after a time, the young polype-bud is produced, and so on for all the zoöid forms subsequently evolved.

The extent to which a coenosarc may be developed varies, however, considerably. It must not be inferred that in every composite Actinozoön such a structure is present; for the mass may exhibit nought else save a congeries of polypes in immediate mutual connection. In this case multiplication by fission or by parietal gemmation has probably occurred (*fig.* 32).

If parietal gemmation be repeated several times during the growth of the budding organism, it is said to be *indefinite*; if once only, *definite*.

Indefinite gemmation is termed *regular* when the young polypes arise at points which are determinate for the same species; *irregular*, when buds are produced indifferently from different parts.

The results of definite gemmation vary according as the producing and produced zoöids are turned towards the same side, or in opposite directions.

Calicular gemmation is only known among certain extinct coralligenous *Actinozoa*. In *Cyathophyllum* and its allies, members of the order *Rugosa*, the primitive polype sends up from its oral disc two or more similar buds, these, in their turn, produce other young polypes, and thus, the process is repeated, until an inverted pyramidal mass of considerable size is formed, all the parts of which rest upon the narrow base of the first budding polype.

Milne Edwards has carefully insisted on the necessity of distinguishing between fission and gemmation among the *Actinozoa*. The oral disc of the budding polype always remains entire, and the bud, when it first appears, wants not only a mouth, but most of the other structures which subsequently it acquires; whereas the polypes produced by fission resemble each other in organisation and, not unfrequently, in size, as soon as they become distinct. Here, it need hardly be said, oral fission is referred to; basal fission, a mere variety of discontinuous gemmation, being, as above stated, of comparatively rare occurrence.

Every polype-bud is, therefore, at first, no more than a protuberance from the two parent layers, enclosing a cæcal diverticulum of the somatic cavity. Thus, the nutrition of the young zoöid is provided for, till it developes for itself a mouth; after which it may either still continue its primitive connection with the common mass or, as already stated, by deposition of tissue secretions, become, physiologically, a separate organism, though morphologically associated with other zoöids of the same composite fabric.

The growth, gemmation, and fission of the composite coralligenous *Actinozoa* require, in addition to what has been said, the following more particular explanations.

The whole compound structure in the sclerobasic species may be regarded as the result of a peculiar modification of the process of basal gemmation; the rudimentary cœnosarc developed around the base of the primitive polype, instead of spreading only at its circumference, shaping itself into a slender and, at first, slightly elevated stem which, gradually increasing in height, continues, at the same time, to excrete a succession of epidermic layers. Thus the young cœnosarc enlarges in diameter, and soon a number of buds are seen to spring from its surface. Some sclerobasic Corals remain thus throughout life, the cœnosarc, with its axial excretion, merely becoming taller. But, more frequently, branches are sent forth, which resemble, in every essential respect, the stem from which they originated.

These branches may be apparently quite irregular in their arrangement, or they may arise in the same vertical plane from opposite sides of the stem, sometimes uniting one with another by the formation of an intricate network, as in the well-known and beautiful Fan-Corals.

Such is the growth of the corallum in the fixed *Gorgonidæ* and *Antipathidæ*. Among the *Pennatulidæ*, which are free forms, the proximal end of the cœnosarc usually becomes produced into a gently swelling or tapering mass, supported by a comparatively slender, elongated sclerobasis.

More varied are the modifications of the composite sclerodermic corallum, the increase of which may be the result of fission alone, or of gemmation alone, or of gemmation and fission combined.

Among coralligenous polypes fission is usually effected by oral cleavage, as indicated above; the proximal extremity of each, for a greater or less extent, remaining undivided. The two zooids thus produced either supply by reparation those parts which were wanting to render them complete, or, it may be, continue through life in a more or less imperfect condition.

The coral structures which result from a repetition of the fissiparous process are of two principal forms, according as they tend most to increase in a vertical or horizontal direction. In the first of these cases the corallum is *cæspitose*, or tufted, convex on its distal aspect, and resolvable into a succession of short diverging pairs of branches, each resulting from the division of a single corallite. In some parts of the mass the walls of the corallites blend, so that a few of them may even become compacted with one another.

When growth takes place chiefly in a horizontal direction, the so-called *lamellar* form of corallum results. Here, the secondary corallites are united throughout their whole height, and disposed in a linear series, the entire mass presenting one continuous theca. Sometimes it is possible to count the number of corallites, but often their several calices merge, as it were, into a single groove, traversed, perhaps, by a columella running parallel to its sides, towards which two opposite rows of septa are seen to converge.

Both the lamellar and cœspitose forms of corallum are liable to become *massive* by the union of several rows or tufts of corallites throughout the whole or a portion of their height. An illustration of this is afforded by the large gyrate corallum of *Mœandrina*, over the surface of whose spheroidal mass the calicine region of the combined corallites winds in so complex a manner as at once to suggest that resemblance to the convolutions of the brain which its popular name of Brain-stone Coral has been devised to indicate.

Basal gemmation, among sclerodermic Corals, affords very different products, according as the cœnosarc remains soft, or deposits a cœnenchyma; appears under the form of stolons, or of stouter connecting stems; or even spreads out in several directions as a continuous horizontal expansion. In this case it is evident that the latest formed parts of the mass are those which are situated nearest to its circumference.

The corallites may either appear widely separated from one another, or closely aggregated and, perhaps, confused by reason of the scanty development of an intervening cœnenchyma.

In some cases the basal deposit encroaches but little on the sides of the corallites; in others it rises upwards till on a level with their calices, or even above them. Again, instead of remaining horizontal, it may become folded in such a manner

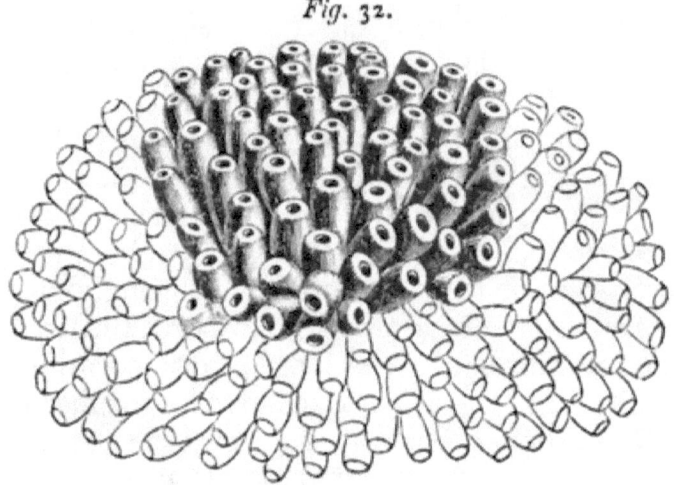

Fig. 32.

TURBINARIA PALIFERA. Corallum, of the natural size.

that one part of its distal surface is brought into more or less close contact with the other; the corallites, which previously seemed to arise from the same plane, having now their calices turned in opposite directions. The resulting corallum will be dense or foliaceous, according to the degree of development of the included cœnenchyma, and the greater or less elongation of its projecting corallites.

Budding from the sides of the corallites takes place, however, more commonly than basal gemmation, not that the latter must be considered of infrequent occurrence.

The aspect of the corallum to which parietal gemmation gives rise varies in accordance with

1. The number of buds which each corallite produces:
2. The frequency with which the budding process is repeated; whether once only or several times during the life of the corallite, but at different stages of its growth:

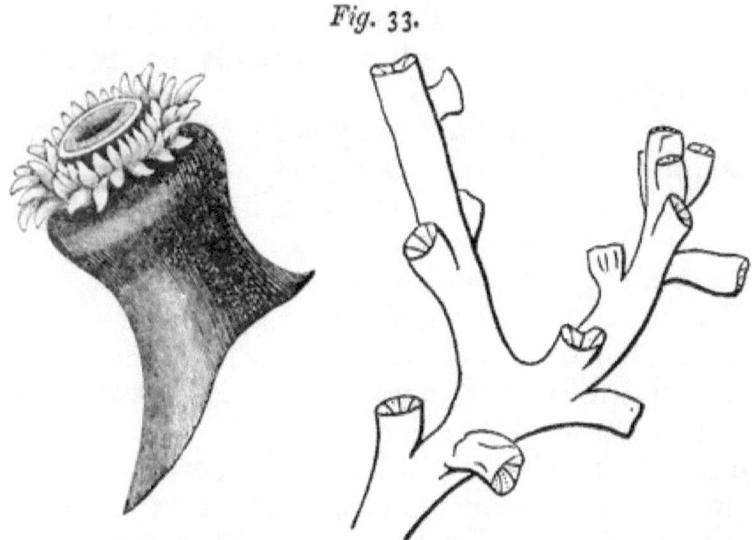

Fig. 33.

DENDROPHYLLIA NIGRESCENS:—Part of a branch, in outline, of the natural size. A single polype is shown separately, magnified.

3. The height which each corallite attains:
4. The angle at which the growing buds diverge from the parent corallite:
5. The degree of rapidity with which such divergence takes place:
6. The position of the buds produced; from one side only of the budding corallite, or indifferently from any part of its circumference; at a short distance from its base, or relatively nearer to the edge of the calice:

7. The degree and nature of the union between the several corallites; whether this be produced by the close contact and blending of their walls, or by the development, both in a horizontal and vertical direction, of an epitheca, cœnenchyma, and other similar structures (*fig.* 33):

and several less essential modifications of the same process, manifested either separately or in combination.

It is, therefore, not surprising to find among budding Corals forms analogous in every respect to those produced by fission and, in addition, many others whose physiognomy, copious as is the list of descriptive terms, it is scarcely possible to define. Nor does each Coral always restrict itself to a single mode of growth, but, on the contrary, several kinds of gemmation, or of gemmation and fission in unison, have been observed to take place in the same species. So that, even with the aid of figures, and extensive suites of museum specimens, the development of the composite Corals can receive illustration in but a limited and imperfect degree. For these organisms, like all others, rightly to be understood, must be studied amid "the glorious variety of Nature" itself, "living and multiplying in their destined homes and habitats."

Mr. Dana, who devoted much time to the examination of the Corals of the Pacific, thus endeavours to describe some of their general diversities of form:

"Trees of coral are well known; and although not emulating in size the oaks of our forests,— for they do not exceed six or eight feet in height,

— they are gracefully branched, and the whole surface blooms with coral polyps in place of leaves and flowers. Shrubbery, tufts of rushes, beds of pinks, and feathery mosses, are most exactly imitated. Many species spread out in broad leaves or folia, and resemble some large-leaved plant just unfolding; when alive, the surface of each leaf is covered with polyp flowers. The cactus, the lichen clinging to the rock, and the fungus in all its varieties, have their numerous representatives. Besides these forms imitating vegetation, there are gracefully modelled vases, some of which are three or four feet in diameter, made up of a net-work of branches and branchlets and sprigs of flowers. There are also solid coral hemispheres like domes among the vases and shrubbery, occasionally ten or even twenty feet in diameter, whose symmetrical surface is gorgeously decked with polyp-stars of purple and emerald green."

Under such aspects appear the living organisms whose combined efforts have mainly constructed those reefs and islands of Coral origin which now lie scattered far and wide over the surface of the tropical ocean. Three principal forms under which such reefs occur have been distinguished by the names of Fringing-reefs, Barrier-reefs, and Atolls.

Fringing-reefs skirt the shores of favourably situated lands, towards which, at a gentle slope, they incline, ending abruptly seawards, where soundings reveal a depth of from 20 to 30 fathoms. The surface of the reef is covered at high water, and forms a nearly level platform, from a few feet to more than a mile in breadth, according to the degree of inclination which the land presents. Its outer margin may be rendered sinuous by bays, or the continuity of the whole reef completely interrupted by one or more irregular inlets.

Barrier-reefs differ from those just mentioned in occurring at a greater distance from shore, a wide channel of relatively smooth and shallow water flowing between, within which terraces of fringing corals sometimes find their proper environment. Outside the reef depths almost unfathomable have been obtained, as close as it is possible to venture amid the rolling surf which breaks in unceasing billows against its surface.

Besides the larger Barrier-reefs which, in certain cases, attain a length of many hundred miles, there are others, in all respects similar, but more variable in size, which are often found surrounding the smaller islands of the Pacific.

From such reefs to Atolls we may trace every possible transition. An Atoll differs from an encircling Barrier-reef in enclosing, instead of a central island with its intervening channel, an uninterrupted surface of calm water, or lagoon. The low-lying strip of land separating this lagoon from the line of white breakers which marks the outer boundary of the Atoll is seldom more than half-a-mile in breadth, and presents an imperfectly circular or prolonged crescentic form, though occasionally, as in Whitsunday Island, the circle is complete.

Reefs, in general, grow only along their outer edge and, when upraised above the sea-level, always appear highest to the windward side.

Clear sea-water, well aërated, at a certain temperature, and a depth of not more than 25 or 30 fathoms, are among the most important of those external conditions which seem favourable, if not essential, to the growth of reef-building Corals.

But to what other influences do the above three classes of reefs, which present so much in common, owe their occurrence? It has been said that they are of Coral origin; yet how is it that some of them rise from depths so considerable, seeing that those living Corals, by which they have been constructed, build only in seas comparatively shallow?

It is not our business here to discuss the various speculative views which have from time to time been put forward on the subject of Coral-formations. Let it suffice to say that Mr. Darwin's theory of elevation and subsidence offers the only consistent explanation of most of their known phenomena which science is prepared to receive.

If we suppose a Fringing reef, together with the area which it surrounds, to sink at a rate not more rapid than the upward growth of its constituent Corals, the reef itself will undergo little apparent alteration, while a channel of water, gradually increasing in width, will appear between it and the more elevated regions of the slowly submerging land. Thus a Barrier-reef is formed. Depression still going on, the land encircled by the reef is reduced to one or more projecting peaks, as in those islands of the Pacific to which allusion has been made. Further subsidence causes these peaks to disappear beneath the sea-level, and the Barrier-reef changes into an Atoll.

Fringing-reefs, therefore, show that the shores which they skirt are stationary or rising, while

Atolls and Barrier-reefs attest that subsidence has taken place.

That such slow subsidence does occur over many parts of the extensive area occupied by Coral-reefs is proved by a number of considerations; some of which were forcibly suggested by Sir Charles Lyell, several years before Mr. Darwin's theory was promulgated.

Observations and experiments made with a view to ascertain the rate of growth of the reef-building Corals have not hitherto yielded a sufficient number of accurately recorded results. That, in certain instances, their growth is rapid, varying, however, with the species, may be regarded as proven; but it is also far from improbable that there are many species which have the same average rate of increase. It may likewise be conceded that the growth of the same species varies at different periods and under different external conditions. In future investigations on this subject the particular form of the corallum, and its mode of fission or budding, should in each case receive attention. For it is obvious that the same amount of calcareous deposit, if appropriated respectively by a massive and by a dendroid species, would give rise to apparently dissimilar quantitative products.

The Coral-polypes are, however, powerless to raise their structures higher than the line of low-water. To effect this, various agencies, but chiefly the forces of wind and ocean, acting upon masses detached from the reef and other marine débris, are brought into play. But the mode of operation of these agencies, the general theory of elevation and subsidence, the conversion of the irregular surface of the reef into one continuous level, and the alterations to which its dead and deeply-submerged portions become exposed in the lapse of time; these, and other kindred subjects of inquiry, fall rightly within the province of the geologist. As monuments of past change, Coral-reefs form the basis of some of the most "splendid generalisations" which his science has deduced. For him an island occupied each region where now without interruption flow the quiet waters of a lagoon. And seas must have once rolled over those existing continents amid whose mountain-chains remains of ancient Coral-reefs abound.

Yet the zoölogist, in taking leave of his own department of this subject, cannot, without satisfaction, contemplate how large a portion of the earth's substance must, during the long lapse of geological time, have formed part of the organised structures of a group of beings who still continue to fulfil, with no impairment of efficiency, the great task for ages allotted them in the scheme of universal nature. Not matter only, but force, he sees made subject to their sway. The physical agencies, which seemed at first to threaten destruction to the growing Coral, are soon successfully overcome, and then pressed into its service. Nay, without their aid, so much of the reef as rises above the ocean level, forming the abode of plants and animals, and finally of man, could not even have existed. But had Coral-polypes not previously laboured, the same forces would have been potent only to destroy.

Section III.

CLASSIFICATION OF ACTINOZOA.

1. Classification.— 2. Order 1: Zoantharia.— 3. Order 2: Alcyonaria.— 4. Order 3: Rugosa.— 5. Order 4: Ctenophora.

1. Classification.—The class *Actinozoa* may be divided into the four orders here defined:

1. *Zoantharia.*—*Actinozoa*, in which the tentacles are simple or variously modified, in general numerous and, together with the mesenteries, disposed in multiples of five or six. Corallum absent or sclerobasic, in most sclerodermic, the septa of each corallite following the numerical law of the soft parts.
2. *Alcyonaria.*—*Actinozoa*, in which each polype is furnished with eight pinnately fringed tentacles. Mesenteries and somatic chambers in number some multiple of four. Corallum sclerobasic or spicular, rarely thecal, and never presenting traces of septa.
3. *Rugosa.*—*Actinozoa*, presenting a sclerodermic corallum, with tabulæ and well developed septa arranged in multiples of four. Soft parts unknown.
4. *Ctenophora.*—Transparent, oceanic, delicate gelatinous *Actinozoa*, swimming by means of ctenophores, or parallel rows of cilia disposed in comb-like plates. No corallum.

2. Order 1: Zoantharia.—The chief modifications of the plan of structure proper to the

Zoantharia have reference to the form of the polype, with its tentacles and corallum, and that of the compound mass resulting therefrom by a repetition of the process of gemmation. The varieties which these animals present in size, colour, and the superficial aspect of the body are too multifarious to admit of any precise general enunciation.

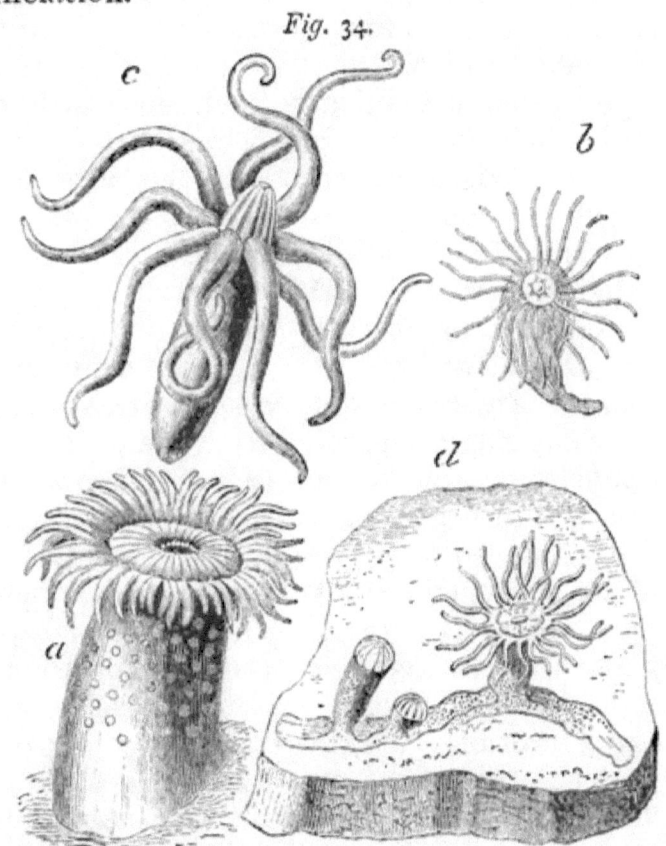

Fig. 34.

ZOANTHARIA MALACODERMATA:— *a*, *Actinia* (or *Sagartia*) *rosea*; *b*, *Ilyanthus Scoticus*; *c*, *Arachnactis albida*; *d*, *Zoanthus Couchii*. (Natural size.)

The form and structure of the polype has been best studied in the soft-bodied Zoantharians, more familiarly known as Sea-anemones. A typical Sea-anemone, when contracted, may be compared to a more or less depressed cone, rounded above, with a gently spreading base; when expanded, to a column, for the most part cylindrical, but widening somewhat towards its extremities (*fig.* 34, *a*). The base is scarcely broader than the column in some species; in *Adamsia*, on the other hand, it is ovate in form, sending forth two lateral lobes, which extend so far as to surround the aperture of the univalve shell on which this curious animal-flower is found, the lobes at length uniting by a zigzag suture along the outer lip of the shell. As in the case of *Hydractinia*, the shells which *Adamsia* selects appear always to be tenanted by a species of Hermit-crab.

The column may have its surface marked by a variety of epidermic growths, or pierced by sundry apertures. In *Actinoloba* its summit rises into a conspicuous ridge, separated from the outer series of tentacles by a deep depression or furrow. Occasionally, as in the same genus, the margin of the disc is waved or thrown into sinuous folds, so that the arrangement of the tentacles appears thereby somewhat confused. The peristomial space may also vary in size, and relative depression or elevation. The lips of the mouth undergo their own modifications, and, in some genera, a groove with mouth-tubercles occurs at but one of the oral angles. In *Actinopsis* these tubercles are produced upwards to form a pair of long, rigid, semi-cylinders, the lateral margins of which again bend downwards to terminate in cleft extremities.

In the non-adherent Sea-anemones, such as *Ilyanthus* and its allies, the column is proportionally larger than in most of the *Actinidæ* proper. The base is either rounded or bluntly tapering, and, in certain genera, becomes at times much distended, as in *Saccanthus* or *Edwardsia*.

In some the base is furnished with a central perforation: in others this appears to be wanting. In *Peachia* the oral region is singularly modified; the tubercles of its single groove uniting to form a tube, the expanded summit of which, 'conchula' of Mr. Gosse, presents a more or less thickened, everted edge, cleft into a variable number of lobes.

The polypes of the *Zoanthidæ* and true coralligenous *Zoantharia*, save in characters merely generic, resemble those of the *Actinidæ*. Their average size is, perhaps, smaller, though the *Actinia Paumotensis* of the Pacific, whose expanded disc measures a full foot in diameter, is, in this respect, certainly exceeded by large specimens of *Fungia*. This genus presents a widely extended, circular or elliptic disc, destitute of the usual folding margin, and blending, by insensible degrees, with the shallow, ill-defined column, over the radiating septa of which the tensely stretched soft parts converge towards the prominent, central mouth. Such a simple form contrasts strikingly with *Meandrina* and those allied genera in which the several polypes produced by fission fuse together into a convoluted linear track, with tentacles arising from its opposite sides.

Of these appendages and their variations, a brief notice seems here required. In *Anthea* (=*Anemonia*) they are long and slender; in *Fungia* and *Discosoma*, reduced to mere warts or papillæ; in *Capnea*, very short, resembling oblong tubercles; while in *Arachnactis*, the outer tentacles are capable of being extended to three or four times the length of the body (*fig.* 34, *c*). In *Eumenides* the tentacles are fusiform, in *Heteractis* moniliform, in *Corynactis* and *Caryophyllia* they terminate in globose heads. Their free extremities are often perforate, the animal having the power of opening or closing the orifice at pleasure. Dana describes the inner tentacles of his *Actinia flagellifera* as terminating in a retractile pencil of hairs, but it is possible that these hairs may have been, in reality, large everted thread-cells. Although in most *Zoantharia* the tentacles are simple, yet in *Thalassianthus* and its allies they are branched, and have their surface studded with tubercles or papillæ. In a few genera, two kinds of tentacles appear on the same polype; the one simple, the other lobed or branched, as in *Phyllactis*.

The number of the tentacles, though in general some multiple of five or six, is, in other respects, liable to considerable variation. Their arrangement, also, is correspondingly diversified. They may dispose themselves in one, two, or more concentric series, and in some species they appear irregularly scattered. *Antipathes* exhibits six tentacles in a single circlet; *Peachia* twelve, similarly disposed; while in the common Sea-anemone there are nearly two hundred of these organs, arranged in the manner noted above. In this species, as in most other *Zoantharia*, the tentacles of contiguous rows alternate. *Cerianthus* and *Saccanthus*, however, possess two distinct circlets of tentacles, the one oral, arising close round the mouth, the other marginal, not far from the edge of the disc, the tentacles of the inner row being equal in number and opposite to those of the outer.

The tentacles of most *Zoantharia* are retractile, but in *Cerianthus*, *Anthea*, and a few other forms, this power is either absent, or imperfectly exercised.

The numerous families of the present order have been conveniently arranged by Milne Edwards under three sub-orders: *Malacodermata; Sclerobasica;* and *Sclerodermata.*

In the *Z. Malacodermata*, the corallum is either absent or represented by scattered spicules. The actinosoma, in most of these animals, presents but a single polype. Exceptions to this rule occur, however, among the *Zoanthidæ*, the budded polypes of which remain permanently united by a cœnosarc, in some linear, in others carpet-like or incrusting. In certain *Actinidæ* also, for example, the *Corynactis mediterranea* of Sars, a similar connection is maintained (*fig.* 34, *d*).

Within the soft parts of the *Z. sclerobasica* spicular tissue secretions seem wanting (*fig.* 35). All the members of this sub-order are composite structures. *Antipathes*, the type of the group, presents a stem-like, simple or branched cœnosarc, which in one species tapers to a length of more than nine feet, with a basal diameter of scarcely ·3 of an inch. In this genus the sclerobasis is horny, and each polype, according to Dana, has but six tentacles; but in the allied family of *Hyalochœtidæ*, the tentacles are twenty in number, while the basal excretion resolves itself into numerous siliceous threads, transparent, twisted into an erect axis. Doubts, however, are yet entertained of the true nature of these so-called

"Glass-plants," whose siliceous stem may be the product, not of the polype-mass, but of the sponge on which it is parasitic.

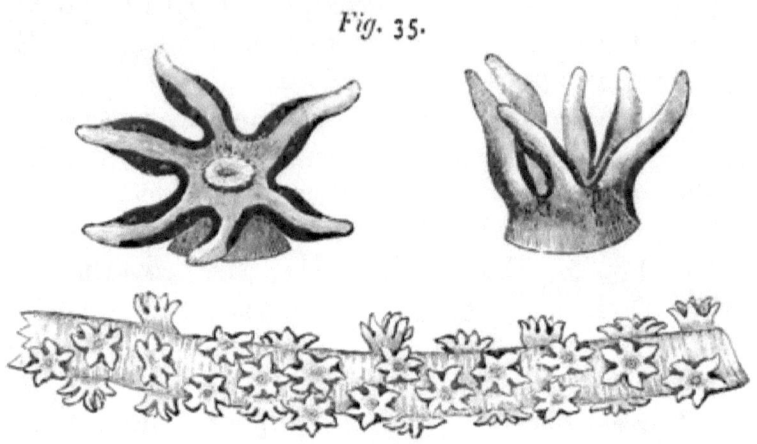

Fig. 35.

ZOANTHARIA SCLEROBASICA:—Part of a living stem of *Antipathes anguina*, of the natural size. Two polypes are shown separately, magnified.

Of the Madrepores, or *Z. Sclerodermata*, want of space forbids us to say much. A short sketch of the several families into which the sub-order is divided appears to be the best form into which this part of our subject may be condensed.

Among the *Turbinolidæ* the corallum is usually simple, never presenting a cœnenchyma. In *Cœnocyathus* continuous lateral gemmation takes place, the corallites so formed remaining connected in a close irregular tuft. In *Blastotrochus* gemmation also occurs, but here it is discontinuous. The septa are very perfectly developed, not giving rise to either dissepiments or synapticulæ (*fig*. 36 *a*).

In the *Oculinidæ* there is a very abundant cœnenchyma, which blends gradually with the thecæ through their entire height. Each corallite has its chambers slightly interrupted by a few dissepiments.

The members of another family, *Astræidæ*, are either simple or composite, but there is no proper cœnenchyma, as in the *Oculinidæ*. The epithecæ or costæ form the chief substance of the mass by which, sometimes, the corallites are separated. Many *Astræidæ* present that rapid fissive development of the corallum, whose curious results have been already indicated in the case of *Meandrina*. Dissepiments, in most, are numerous. In number of genera and species, perhaps, also, of individuals, and, apparently, in general importance, the *Astræidæ* seem to surpass all other families of the class *Actinozoa*.

The *Fungidæ* are at once distinguished from the three families just noticed by the possession of synapticulæ. Dissepiments are absent, nor can it be said in many cases that a proper theca exists; the septa passing, without interruption, into the costæ, save at the base of each corallite. Both simple and composite *Fungidæ* occur, the latter multiplying by lateral gemmation.

In the somewhat porous condition of their ill-developed theca the *Fungidæ* may be said to differ from other Aporose *Zoantharia*, and approach the two next families, collectively distinguished by the title of *Perforata*. These, like the *Aporosa*, have a well-marked septal apparatus, and present no traces of tabulæ.

Among the *Madreporidæ* the sclerenchyma is simply porous, the septa are distinct, and but very slightly perforate. Save in *Eupsammia* and some of its allies, the corallum is composite, but the thecæ of the separate corallites do not become lost in the surrounding cœnenchyma. In the *Poritidæ*, on the contrary, such fusion always takes place, and the septal system, instead of forming distinct plates, consists wholly of more or less definite series of trabiculæ. The entire corallum, in like manner, appears to be made up of a spongy, reticulate sclerenchyma.

The next division, *Tubulosa*, contains only a single family, *Auloporidæ*; and this but two

Fig. 36.

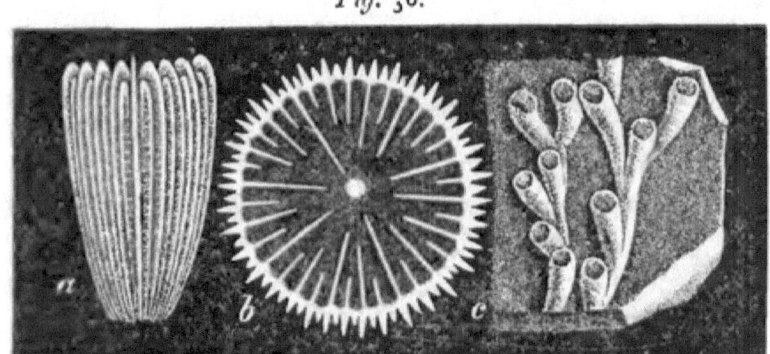

ZOANTHARIA SCLERODERMATA:— *a*, corallum of *Turbinolia costata*; *b*, the same, in transverse section, showing the columella, septa, theca, and costæ; *c*, part of corallum of *Aulopora tubæformis*. (All, except *c*, magnified).

genera, *Pyrgia* and *Aulopora*, in both of which the corallite, while destitute of tabulæ, has its septal system indicated by faint markings along the inner surface of a comparatively smooth tube. *Pyrgia* is simple, though having a distinct epitheca. In *Aulopora* the somewhat remote corallites are connected by means of a basal creeping cœnenchyma. (*fig.* 36, *c*.)

The four last families of *Zoantharia* constitute

the great division of *Tabulata*, in which the rudimentary condition of the septa is made amends for by the extensive development of the transverse floors, referred to in the name this group bears. (*fig.* 29.)

The corallum of the *Tabulata* is mostly, if not always, composite. Among the *Milleporidæ* an abundant cœnenchyma occurs, and the resulting compound structure assumes a massive or foliaceous aspect. The substance of the corallum is traversed by interspaces which give to its section a somewhat tubular or cellular appearance. In the *Seriatoporidæ* it is more compact, but here, likewise, the cœnenchyma is abundant, presenting, externally, a tufted or arborescent form. In the *Favositidæ* the corallites have their lamellar walls brought into very close apposition, little or no true cœnenchyma being observable; while in the *Thecidæ* the septa form by their lateral union the greater portion of the dense spurious cœnenchyma, of which their massive corallum is composed.

The following may be given as definitions of the families of *Zoantharia*. Those of the Madreporic forms, founded wholly on characters derived from the corallum, admit readily of being exhibited under the guise of an analytical table.

Order ZOANTHARIA.

Sub-order 1. *Z. Malacodermata.*

Family 1. ACTINIDÆ.

Corallum not evident. *Polypes* rarely connected by a cœnosarc; in general, locomotive, the flattened base of each adhering at pleasure.

Family 2. ILYANTHIDÆ.
Corallum not evident. *Polypes* unattached, with rounded or tapering base; no connecting cœnosarc.

Family 3. ZOANTHIDÆ.
Polypes attached, united by a cœnosarc, and furnished with a spicular *corallum*.

Sub-order 2. *Z. Sclerobasica.*

Family 4. ANTIPATHIDÆ.
Corallum sclerobasic, horny, smooth or spinulous. *Polypes* with six tentacles.

Family 5. HYALOCHŒTIDÆ.
Corallum sclerobasic, composed of twisted siliceous fibres. *Polypes* with twenty tentacles.

Sub-order 3. *Z. Sclerodermata.*

1	Tabulæ present. Septa rudimentary (*Tabulata*)..		2
	Tabulæ absent.		5
2	Cœnenchyma wanting, or ill-developed..		3
	Cœnenchyma abundant.		4
3	Septa forming a spurious, massive cœnenchyma.	Family 6. THECIDÆ.	
	Septa and corallites distinct..	Family 7. FAVOSITIDÆ.	
4	Sclerenchyma compact. Corallum arborescent.	Family 8. SERIATOPORIDÆ.	
	Sclerenchyma tubular or cellular.	Family 9. MILLEPORIDÆ.	
5	Septa indicated by faint striæ (*Tubulosa*).	Family 10. AULOPORIDÆ.	
	Septa well developed.		6
6	Sclerenchyma porous (*Perforata*)		7
	Sclerenchyma imperforate (*Aporosa*)..		8

7	Sclerenchyma reticulate. Thecæ not distinct from the surrounding cœnenchyma.	Family 11. PORITIDÆ.
	Sclerenchyma simply porous. Thecæ distinct.	Family 12. MADREPORIDÆ.
8	Synapticulæ present. No dissepiments.	Family 13. FUNGIDÆ.
	Synapticulæ absent.	. . 9
9	Dissepiments in general numerous. Cœnenchyma absent, or formed only by the development of the costæ or epitheca.	Family 14. ASTRÆIDÆ.
	Dissepiments few or absent.	. . 10
10	Cœnenchyma compact, abundant.	Family 15. OCULINIDÆ.
	No cœnenchyma.	Family 16. TURBINOLIDÆ.

In addition to those here defined, Milne Edwards has distinguished four other families of *Aporosa*, which inosculate, so to speak, between the primary groups just mentioned. The first family, *Dasmidæ*, includes but a single genus, closely related to the *Turbinolidæ*, from which it differs in the peculiar modifications of its septa. Each of these is represented by three vertical laminæ, united only along their external margin. A second family, *Stylophoridæ*, appears as a transitional group between the *Oculinidæ* and *Astræidæ*. As in the former, there is a well-developed cœnenchyma and few dissepiments; but, on the other hand, the surface of the cœnenchyma is echinulate, while it is smooth in the *Oculinidæ*, and the thecæ of the corallites do not, as in that family, increase endogenously, so as almost to obliterate the loculi. Another osculant family, *Echinoporidæ*, still more closely resembles the *Astræidæ*, differing therefrom chiefly in the possession of a foliaceous, basal cœnenchyma. But one genus presenting this combination of characters has been observed. The family *Merulinidæ* has, lastly, been constituted for the reception of an equally aberrant Astræoid genus, *Merulina*, which clearly points in the direction of the *Fungidæ*, resembling these corals in the perforate condition of its corallum, though, as in the true *Astræidæ*, no synapticulæ occur.

3. **Order 2 : Alcyonaria.** — The *Alcyonaria*, with the exception of one genus, *Haimeia*, which may, however, yet prove to be an immature form, are composite in structure; their polypes being mutually connected by a cœnosarc, through which permeate prolongations of the somatic cavity of each, forming a sort of canal system, whose several parts freely communicate and are, therefore, readily distensible.

Throughout the whole order the polypes exhibit a very close agreement in structure, howsoever much the cœnosarc may vary. Each, when expanded, displays a cylindrical, or somewhat octagonal, tube, with delicate transparent walls, and eight pinnate tentacula, whose form offers slight though characteristic variations among the several genera of the group. In some the polypes are retractile into excavations which occur in the substance of the cœnosarc, while in others such excavations seem to be wanting.

Alcyonium, the typical genus, presents, when first dredged up, a sufficiently repulsive aspect, suggestive of the vulgar names, "Cow's paps" and "Dead-man's hand," sometimes conferred on it. But, when placed in sea-water, the lobate fleshy mass, distending its aquiferous system, is gradually seen to become exquisitely pellucid, while from all parts of its surface numbers of tiny polypes, emerging, expand to the utmost their star-shaped crowns of delicately fringed tentacula. Within the somatic chambers circulating currents may now be observed. These find their way up one side of the tentacles, following the course of the several fringes, and, having gained their summits, again revert, proceeding in a contrary direction. So that here, as in many *Zoantharia*, it would not, perhaps, be too much to say that the tentacles, by reason of the delicacy of their ciliated walls, fulfil the proper function of a respiratory system.

In *Sarcodictyon*, as in *Alcyonium*, a spicular corallum occurs, but the cœnosarc is scanty and creeping, resembling that of the *Zoanthidæ*. So also in *Cornularia*, the corallum of which is, however, more consolidated. But of all the *Alcyonidæ* proper *Telestho*, with its tufted, sub-calcareous, tubular corallum, makes the nearest approach to the allied family of *Tubiporidæ*.

The beautiful Organ-pipe Corals, forming the several species of the genus *Tubipora*, appear to be the sole representatives of this group. Allusion has already been made to the exceptional structure of their corallum, the colour of which, in all cases, is of a bright crimson-red. The polypes are either violet or grass-green in tint, and, according to the dissections of Dana, present this anatomical peculiarity, that two only of the mesenteric edges are furnished with ova, the remaining six supporting spermaria. The oral extremity of each polype can be inverted for protection into the summit of its calcareous tube, but it is wrong to suppose that the latter completely invests the soft parts of the animal, the corallum of *Tubipora* being a true tissue secretion. Its horizontal outer plates are suggestive of a distinct analogy to the *Tabulata*, nor are traces of internal tabulæ wholly wanting. The characteristic form of *Tubipora* seems due to the periodic budding of zoöids from the distal surface of the plates, while at the same time certain of the older corallites continue to increase in height. But neither the minute structure nor development of this interesting genus have yet received proper attention.

The *Gorgonidæ* differ from all other *Alcyonaria* in having an erect branching cœnosarc, firmly rooted by its expanded proximal extremity (*fig.* 37, *c*). Those which possess a horny sclerobasis have been by many writers confounded with the *Antipathidæ*; but, apart from the anatomical features of their polypes, they may at once be known from the latter by the more or less sulcate aspect presented by the surface of the sclerobasis. The modifications which this structure displays in *Corallium*, *Isis*, *Mopsea*, and *Melitæa* have already received a brief notice. It may suffice to add that very exaggerated conceptions seem to prevail as to the height which the horny *Gorgonidæ* are capable of attaining. It is doubtful whether their largest trees ever rise to more than five or six feet, yet some have been reputed to rival oaks in size, an assumption which, however incredible, is, nevertheless, not inconsistent with theoretical considerations.

The *Alcyonaria*, as a group, seem destitute of locomotive power, though one family of this order, the *Pennatulidæ*, have been often regarded as oceanic, and described by such names as "Polypes nageurs." It is more probable, however, that, under ordinary circumstances, these creatures live with their proximal extremity plunged firmly into the sand or mud of the sea-bottom; the distal

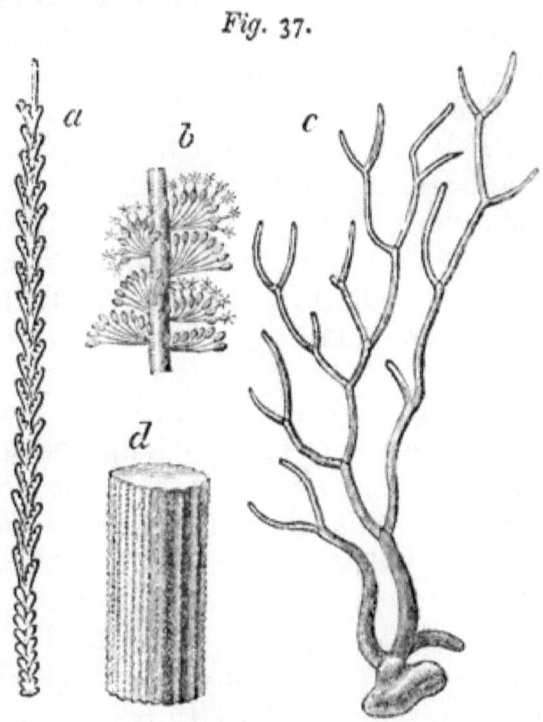

Fig. 37.

PENNATULIDÆ and GORGONIDÆ: — *a*, dried stem of *Virgularia mirabilis*; *b*, portion of another stem, in the living condition; *c*, corallum of *Mopsea costata*; *d*, fragment of the same. (*a* is reduced one-third; *b* and *c* are of the natural size; *d* is magnified.)

end of the cœnosarc, which bears the numerous polypes, freely exposing itself to the influence of the clearer water above.

The cœnosarc of the *Pennatulidæ* may be slender and simply elongate, with very short pinnules, or lateral lobes, bearing the polypes, as in *Virgularia*, the sclerobasis of which is rigid, tapering towards its extremities, and densely calcareous (*fig.* 37, *a*). In the true Sea-pens, forming the genus *Pennatula* and its near allies, the pinnules are very conspicuous, and so modified as to arrangement and comparative size that the whole mass presents a striking resemblance to a bird's feather. The proximal end of the cœnosarc, often for nearly half its length, is bare of pinnules or polypes, appearing swollen and fleshy. In other *Pennatulidæ* the entire cœnosarc is club-shaped, without any pinnules, the polypes being irregularly scattered, as in *Veretillum*, or arranged in longitudinal rows on part only of the surface, as in *Kophobelemnon*. In *Renilla* a comparatively short cœnosarc expands distally to support a smooth, symmetrical, kidney-shaped disc, from the free surface and edge of which the scattered polypes arise. This aberrant genus appears to want a sclerobasis, the interior of the stalk and disc being hollow, and in free communication with the cavities of the polypes, so that the animal possesses the power of largely increasing its dimensions by allowing itself to become expanded by the ingress of the surrounding sea-water.

Like the members of the preceding families, many *Pennatulidæ* are liable to have their soft structures strengthened by the deposition of spicular concretions.

The Sea-pens still further interest us by reason of their beautiful phosphorescence. What Agassiz has observed in *Renilla* is probably true of the entire group. "It shines at night with a golden green light of a most wonderful softness. When excited, it flashes up more intensely, and when suddenly immersed into alcohol, throws out the most brilliant light." Experiments performed on our own British *Pennatula phosphorea* led Professor E. Forbes to the conclusion that this species is "phosphorescent only when irritated by touch"; but it seems safer to infer that the light, itself the index of an energetic display of vital power, is only evoked in answer to proper stimuli, which may very well be expected to occur more appropriately in nature, though, doubtless, under a form less clumsy, than in the exaggerated conditions of an experiment. Forbes also showed that the phosphorescence, when thus excited by shock, sparkles onward from the portion struck in an upward or distal direction, still, however, continuing to be emitted from the point of prime contact. The vividness of the luminosity appears to bear a direct ratio to the living energy of the animal. Such are the chief conditions of its manifestation, but the true cause of this phenomenon, as of vital phosphorescence in general, still remains almost wholly unknown.

Four families of *Alcyonaria* may be defined:

Order ALCYONARIA.

Family 1. ALCYONIDÆ.

Corallum sclerodermic, in general spicular, without true calcareous thecæ. *Cœnosarc* fixed.

Family 2. TUBIPORIDÆ.

Corallum consisting of a number of distinct corallites, destitute of septa, their thecæ united externally by horizontal plates, arranged at distant intervals. *Cœnosarc* fixed.

Family 3. PENNATULIDÆ.

Corallum sclerobasic, tissue secretions also being sometimes present. *Cœnosarc* free.

Family 4. GORGONIDÆ.

Corallum sclerobasic, sulcate, with or without additional tissue secretions. *Cœnosarc* shrub-like, attached by its expanded proximal extremity.

4. **Order 3: Rugosa.**— Among the *Rugosa* a highly developed sclerodermic skeleton occurs, each corallite being very distinct, and presenting, in many cases, both septa and tabulæ. Some *Rugosa* are simple, the corallite often attaining a considerable size; others are composite, increasing either by lateral or calicular gemmation, these two processes, but especially the latter, checking to a greater or less extent the growth of the primitive corallite. A true cœnenchyma is absent.

In *Stauria*, *Holocystis*, and *Polycœlia* four of the septa admit readily of being distinguished, by their greater development, from the others, but in *Cyathaxonia* only one primary septum or chamber remains conspicuous. So, also, of *Zaphrentis* and its more immediate allies (*fig.* '38). In other genera the septa radiate, in about an equal manner, from the inner surface of the theca. In many *Rugosa* they are incomplete, that is, "do not extend from the bottom to the top of the corallite, in the form of uninterrupted laminæ."

The tabulæ exhibit various grades of development, and, in some species, are wanting altogether.

In a few *Rugosa* the columella is cylindrical, and of large size; in others, styliform or lamellar:

often, it is wanting. In *Cyathophyllum* and certain allied forms there is a spurious columella formed by the "twisting together" of the inner edges of the septa.

In *Cystiphyllum* neither columella, septa, nor tabulæ can be distinguished. The whole interior of the corallite, save its shallow calice, is here, as

Fig. 58.

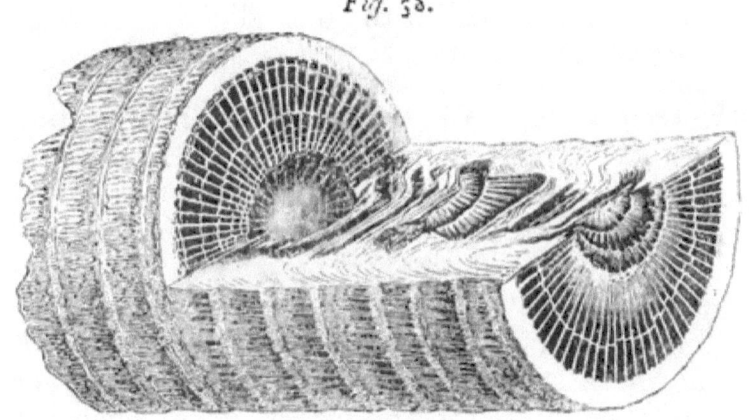

ZAPHRENTIS CYLINDRICA.
(Part of a corallite, of the natural size.)

it were, broken up by the numerous laminæ of a sclerenchymatous deposit resembling the vesicular substance of its epitheca and wall.

Most *Rugosa* belong to the large group of *Cyathophyllidæ*. The remaining families include but a few generic forms.

Order RUGOSA.

Family I. STAURIDÆ.
Corallum simple or composite. *Septa* complete, united by lamellar dissepiments.

Family 2. CYATHAXONIDÆ.
Corallum simple. *Septa* complete. No dissepiments or tabulæ.

Family 3. CYATHOPHYLLIDÆ.
Corallum simple or composite. *Septa* incomplete. Tabulæ, in general, present.

Family 4. CYSTIPHYLLIDÆ.
Corallum simple; composed chiefly of a vesicular mass, with but slight traces of *septa*.

5. **Order** 4: **Ctenophora.**— The leading characters of the *Ctenophora*, or oceanic *Actinozoa*, have been already, to some extent, sketched out in the account given of *Pleurobrachia*, selected as the type of the group, and in the comparative survey which has been taken of the principal organic systems of the present class.

The more striking modifications which appear within the limits of the order have reference chiefly to size, the general form of the body, the several parts of the canal system, and the structure and arrangement of the tentacles.

The ordinary dimensions of *Pleurobrachia* are by many other *Ctenophora* frequently exceeded, a common British species of *Bolina* often attaining a long diameter of two or three inches, while *Beroe* sometimes reaches the size of a large lemon.

In *Pleurobrachia* alone does the form of the body approach the spheroidal (*fig.* 39, *e*). Its axis is somewhat lengthened in the *Beroidæ*, so that the animal, when seen in profile, appears more or less ovate (*c*). In *Cestum*, on the other hand, elongation takes place to an extraordinary extent, at right angles to the direction of the digestive track, a flat ribbon-shaped body, three or four feet in length, being the result (*a*). *Calli-*

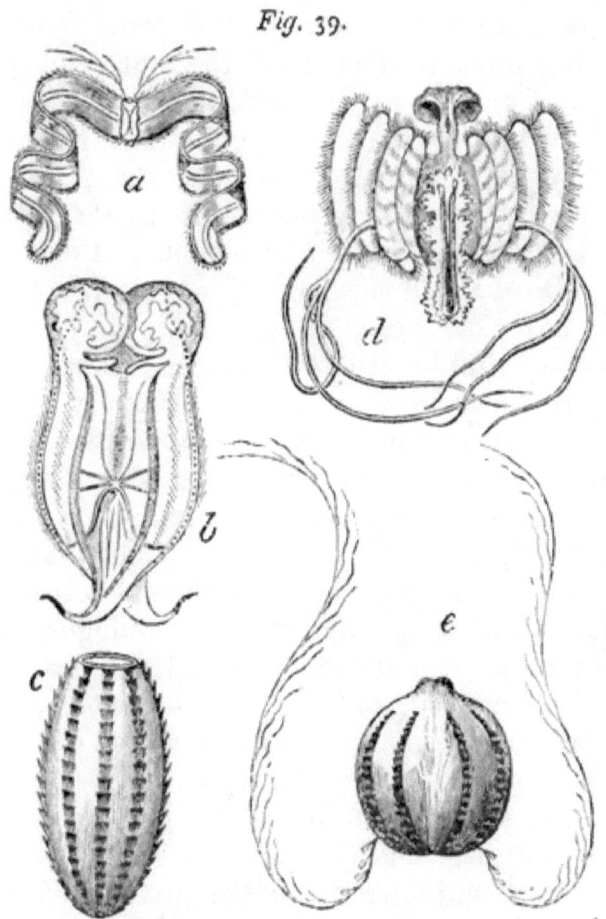

Fig. 39.

Various forms of CTENOPHORA:— *a*, *Cestum Veneris*; *b*, *Eurhamphæa vexilligera*; *c*, *Beroe rufescens*; *d*, *Callianira triloptera*; *e*, *Pleurobrachia pileus*. (*a* is considerably reduced; *b*, slightly so; *c* and *e* are about the natural size: the size of *d* is uncertain.)

anira, a genus of which very little is known, is remarkable for having its ctenophores elevated on prominent wing-like appendages, and may possibly yet prove to be but an imperfectly developed member of the group (*d*).

In one large division of the order two wide diverging lobes project from the antero-posterior regions of the body, far beyond the level of the mouth, which, by their approximation, they sometimes wholly conceal. Between these lobes, in *Bolina* and its allies, occur on either side two small lateral appendages, or earlets. Four of the ctenophores, much shorter than their fellows, run to the bases of these, along which they are continued as simple ciliated fringes. The earlets, therefore, may be regarded as specially modified prolongations of the lateral ctenophores.

Eurhamphœa, one of these lobed *Ctenophora*, is remarkable for the general elongation of the axis of the body, which is, moreover, much compressed from side to side (*b*). The two lateral actinomeres terminate in long tapering appendages, which project for some distance beyond the apical extremity of the animal, and then curve gently upwards and outwards. Of a similar nature, but much shorter and wider, are the apical appendages, or lappets, of some *Beroidæ*.

The mouth of the *Ctenophora* varies as to size, degree of prominence, and the relative development of distinct lips. Even in the same individual its form, at different periods, presents many diversities of aspect. It attains a very large size in *Beroe* and its allies, extending right across almost the entire oral extremity, in its usual antero-posterior direction. Hence Leuckart has proposed to divide the *Ctenophora* into two separate groups; the *Eurystomata*, corresponding to the *Beroidæ,* and the *Stenostomata,* including all remaining members of the order.

The apical area, also, presents its own peculiar variations. In general it appears as a somewhat oblong flattened space, bounded by a distinct ridge, its long diameter coinciding with that of the mouth. The ridge is usually smooth, and the general surface of the included region covered with very fine cilia. But in the *Beroidæ* a number of conspicuous, though short, arborescent filaments fringe the margin of the area, which, in this family, appears divided into a pair of shallow, ovate lobes, converging to a point at the apical pole of the body.

The modifications of the nutrient system now require our attention. The whole of this apparatus may, for purposes of description, be resolved into the following minor systems or groups of parts:—

1. An axial system; including the digestive tube, the funnel, and the apical canals.
2. A paraxial system. To this belong the paragastric canals arising from the funnel, and an oral vessel, or vessels, into which, in some genera, their distal extremities open.
3. A ctenophoral system; the eight canals of which may be variously prolonged to supply the lobes or other appendages of the body; and—
4. A radial system; or those channels whereby the several elements of the preceding system become connected with the funnel.

1. The parts of the axial system vary but little throughout the order. In *Pleurobrachia* the di-

gestive sac is relatively shorter than in other *Ctenophora*. In the *Beroidæ* the funnel is very short and wide. Not more than one pair of apical canals ever appears to have been noticed.

2. Of paragastric canals, one or two pairs may be present. The single pair of these canals are cæcal in *Pleurobrachia*; in *Beroe* they open into an oral tube. In most of the lobed *Ctenophora* two pairs of paragastric canals occur, of which one lies close to the sides of the digestive sac, the position of the second pair being more superficial. The inner pair may be cæcal, while the outer pair open into the oral tube; or the latter may anastomose with both pairs of paragastric canals, as described by Milne Edwards in *Chiajea Palermitana*.

The oral vessel may be circular, as in the *Beroidæ*, or consist of two straight canals running by the sides of the mouth, as in many of the lobed *Ctenophora*; though in one of these, *Eurhamphœa*, this vessel is complete.

3. The eight ctenophoral canals may conveniently be divided into four lateral, and four antero-posterior.

The oral extremities of all these canals may be cæcal, as in *Pleurobrachia*, or open into the circular tube surrounding the mouth, as in the *Beroidæ*. Among the lobed *Ctenophora*, the same tube, or the two vessels its representatives, receives only the extremities of the four lateral canals. Thus, in *Bolina*, each of the latter, after supplying the course of its comparatively short ctenophore, winds round the edge of one of the smaller lobes or earlets, and having again reached the general surface of the body, soon joins the extremity of the oral tube of that side, before gaining which it sends off a branch, destined, after having pursued a long and devious track, to anastomose with its fellow from the opposite side of the body, and the complex system formed by the much convoluted and prolonged extremities of the intervening pair of antero-posterior canals. These last, after quitting their ctenophores, which are relatively much longer than those of the sides of the body, are produced to supply the two large lobes, in front of and behind the mouth, within the substance of which, after many sinuous turns, they finally blend and become lost.

The course of the ctenophoral canals is usually simple. In *Chiajea Palermitana* each gives off on either side a number of very short straight branches, which in the *Beroidæ* are replaced by somewhat larger, arborescent tufts.

4. Lastly, the apical extremities of the ctenophoral canals, and the manner of their communication with the axial system, or, in other words, the disposition of the radial vessels, remain to be noticed.

In *Pleurobrachia*, as has been shown, the primary, secondary, and tertiary radial canals are very well marked, meeting the ctenophoral vessels at right angles to, and about midway in, their course. The apical extremities of the latter are distinctly prolonged, to end cæcally around the oblong, flattened area.

In the *Beroidæ* a radial system can scarcely be said to occur, the apical ends of the ctenophoral canals curving gently round, to open into the funnel, which is placed very close to its own extremity of the body.

Arrangements intermediate between the two extreme conditions just noticed may be traced among the several genera of the lobed *Ctenophora*.

Thus, in *Le Sueuria*, the apical extremities of the eight canals curve inwards to empty themselves into the four short radiating vessels which issue from the funnel. Somewhat similar is the radial system in *Bolina* and, perhaps, in some species of *Chiajea*. But in *C. Palermitana*, while the four longer or antero-posterior canals comport themselves much as in *LeSueuria*, each lateral canal, near the middle of its course, is connected by a short transverse branch with the extremity of one of the four short radial vessels, just where it receives the curved inward prolongation of the adjacent antero-posterior vessel. In this form the apical extremities of the shorter ctenophoral canals are cæcal, but in *Eurhamphœa*, whose radial system resembles that just noticed, the lateral canals of each side open at an acute angle into one another, at the bases of the two curved appendages of the apical lobes peculiar to this genus.

Comparing the preceding account with the little yet known of the development of the nutrient apparatus in the *Ctenophora*, the following conclusions seem deducible. The first formed part of this apparatus seems to be that large rudiment of its axial system from which, at an early period, the digestive sac and funnel become differentiated. From the funnel, or central portion of the whole nutrient cavity, the apical canals soon branch off.

Thus the axial system is completed. Next, radial and paragastric canals appear, the former quickly reaching the surface of the body, where they branch and give origin to the ctenophoral canals, the apical ends of which are well developed, while their opposite extremities are still on their way towards the oral region, in striving to gain which they are outstripped by the paragastric vessels. The extremities of these either remain cæcal (*Pleurobrachia*), or, by branching, give origin to the two lateral rudiments of the oral canal. These may still continue distinct (*Bolina, Chiajea, Le Sueuria*), becoming connected with the extremities of the lateral ctenophoral canals; or unite to form a complete circular tube, which receives, as before, only the four lateral canals (*Eurhamphœa*), or both these and the vessels of the antero-posterior ctenophores (*Beroe*). Thus, the oral tube bears to the paragastric canals a relation comparable, perhaps, with that between the ctenophoral vessels and the radial system: and the gradual development of the entire canal system may be described as tending in a peripheral direction, while its several elements bifurcate; the branches so formed, to a greater or less extent, again uniting, and prolonging their course beneath the surface of the body.

In the above survey the canal systems of *Cestum* and *Callianira* have not been included. Of the latter nothing whatever is known. In *Cestum* the axial system resembles that of *Bolina* and the *Ctenophora* in general. Each half of the ctenophoral system is represented by four very long canals, two of which run side by side along one of the fringed margins of the ribbon-shaped body, the other pair, parallel to these, being situate midway between them and the opposite, or oral, margin. For, in this strangely aberrant form, the course of the axial system corresponds with the short mid-axis of the ribbon, the apparent width of which represents, therefore, the true height of the animal, whose breadth answers to the length of the ribbon, and antero-posterior diameter to its thickness. On this account, it will be convenient to speak of two of the ctenophoral canals as marginal, the two others being medial. There are four radial vessels, two on either side, each of which divides into a single pair of branches, communicating with the ctenophoral canals. The two branches of each radial canal are very unequal in length, and run in opposite directions, the shorter branch soon becoming continuous with a marginal canal, while the longer branch trends parallel to the sides of the digestive cavity, turning round abruptly to open at right angles into one of the medial canals, as soon as it has reached its level. At each extremity of the ribbon the marginal and medial canals anastomose with one another and a long vessel running parallel to their course along the oral margin of the body. At its opposite extremity, near the mouth, this canal unites with its fellow of the other side, where both are joined by a pair of paragastric canals, which Milne Edwards has figured as running in an antero-posterior direction, parallel to the digestive sac. A second, or inner, pair of these canals has not yet been observed.

There can be little doubt that the paired medial and marginal canals of *Cestum* represent the eight ctenophoral vessels of other genera of the order, and that each half of the long unpaired marginal canal is homologous with one of the lateral oral vessels in such genera as *Bolina* or *LeSueuria*.

Two principal kinds of tentacles occur in the *Ctenophora:* long, highly contractile cords, capable of being retracted into special pits; and shorter, isolated threads, which may, in some species, become aggregated to form tufts or bunches. Among the *Beroidæ*, tentacles are absent. In *Pleurobrachia* a large tentacular pit excavates obliquely upwards the substance of the two lateral actinomeres. The base of this pit is brought into close connection with the distal extremity of the short primary radial canal, which opens directly into a wide heart-shaped sac, from between the two deeply-cleft lobes of which, at the upper portion of the fissure formed by their junction, the tentacle itself makes its appearance. Proximally, it is somewhat compressed, but for the greater part of its length becomes truly cylindrical, giving off on that side which is turned away from the body a number of secondary lateral filaments. Both these and the tentacle itself are hollow, communicating with the canal system through the medium of the basal sac, their walls also, like those of the body canals, being lined by an investment of endoderm. On the secondary branches themselves still more minute threads are said to have been observed. Of the grace and beauty which the entire apparatus presents in the living animal, or the marvellous ease and rapidity with which it can be alternately contracted, extended, and bent at an infinite variety of angles, no verbal description can sufficiently treat. These movements seem partly caused by the action of the contractile fibres which occur in the tentacular walls, and partly by the distensive pressure of the fluid forced into the interior of the tentacle, by means of the elastic basal sac.

An account somewhat different to the above has recently been given by Professor Agassiz, both of the precise structure of the tentacle itself, and the mode of its connection with the canal-system. "Nearly two-thirds of the length and breadth of the proximate side of the actinal or closed end of the tentacular socket is occupied by an oblong disc, from the mid-length of which the tentacle arises. The distal side of the disc, or that which faces towards the periphery of the body, is convex, with a shallow furrow extending from the base of the tentacle to the actinal end of the disc; and the proximate side, or that which faces towards the axis of the body, is a plane, immediately beneath whose surface and next to the edge, two chymiferous tubes run parallel; leaving between them, along the median line, a thick ridge, which is nearly as broad as the diameter of the tubes."
—" Only imagine the socket to be removed or reverted, as oftentimes does happen in a great measure, and the whole apparatus will appear like a peripheric ridge, which, at one point, is drawn out into a slender thread, the tentacle. The base of the tentacle has the form of a high, narrow ridge or keel, more or less plicated or distorted, according to whether the apparatus is extended or retracted; but we have never seen it projecting beyond the aperture of the socket. At the basal end of the keel it is as broad as the disc from which it arises, but it suddenly narrows to a uniform thickness, which it retains to the other end, where-

it merges into the cylindrical part of the tentacle." Professor Agassiz describes the latter not as hollow but solid, though he recognizes the two layers of which it is composed.

In *Callianira* two long tentacles, relatively situate as in *Pleurobrachia,* but destitute of lateral threads, divide, at their free extremities, into two or three lengthened branches (*fig.* 39, *d*).

In *Cestum,* also, a pair of symmetrical tentacles appear to be usually present, but these do not, as in the preceding forms, issue from the equatorial region, thence turning away from the mouth; but, rather, take their position in front of and behind the latter, towards which they are seen to incline. Milne Edwards figures the tentacles of *C. Veneris* as simple; by other writers, and perhaps in other species, they are described as variously branched.

Among the lobed *Ctenophora* the particular homologies of the tentacular apparatus have hitherto been by no means sufficiently studied. In *Chiajea* occur two tentacles, one on each side of the body, but similar in other respects to those of *Cestum*. Tentacles of the second type are, however, more frequently to be met with in this section of the order, and these may be either lateral, and arranged in groups, as in most genera, or disposed in a ring round the mouth, as in *Eurhamphœa*.

In *Bolina* a tuft or brush of very short tentacles is seen to arise on either side of the mouth, where the oral vessels appear to meet, about midway in their course, the superficial paragastric canals. The extremity of each of these canals ends in a simple socket, within which the tuft of tentacles may be withdrawn. But there is no proper sac, as in *Pleurobrachia.* Agassiz states that, in this genus, only the innermost pair of paragastric canals open into the oral vessels, the outer pair, notwithstanding their close approximation to the sides of the mouth, being destined solely for the supply of the tentacular bulbs.

In *LeSueuria* two tufts of tentacles, similar in position to those of *Bolina,* but prolonged to form a pair of lateral fringes, were first observed by Milne Edwards, and in addition four simple conical appendages, of moderate length, arise, one between each of the four pairs of smaller lobes characteristic of this genus. The outer paragastric canals are seen distinctly to open into the canals representing the oral vessels, but Milne Edwards does not notice the existence of any communication between them and the lateral tentacular fringes, which are, perhaps, nothing more than filamentous extensions of the ectoderm.

LeSueuria, according to the same writer, is furnished, however, with a pair of curious appendages by means of which the oral extremities of the paraxial canal system communicate directly with the exterior. Each appendage is cylindrical, short, and tubular, arising from the midst of one of the principal tentacular tufts, and terminating distally in four small lobes, surrounding the orifice of the canal, which seems to perforate its axis. Agassiz denies the existence of this opening, and considers the two appendages homologous with the tentacular bulbs of *Bolina.*

In *Leucothea* a very complicated tentacular apparatus occurs; short threads like those of *Eurhamphœa* and *LeSueuria* being here present, in addition to three long tentacular organs, arising on each side of the mouth. Of these, one is simple, while the others display a number or lateral ramifications. But all these structures require to be investigated anew.

In habit the *Ctenophora* resemble the oceanic *Hydrozoa*, like them swimming near the surface in calm weather, and again descending on the approach of showers. Like them, also, their delicate structures rapidly disappear when removed from the sea-water and exposed to the rays of the sun, an almost imperceptible film remaining the only trace of what was erewhile an active and beautiful organism. Yet are the *Ctenophora* very voracious, feeding on a number of floating marine animals, among which their own kindred seem especially to be preferred. The prey, once swallowed, is assimilated with a rapidity which to some may seem strange, when the simple structure of the digestive apparatus is considered. All the *Ctenophora* are not equally fragile. *Pleurobrachia*, in spite of its tender gelatinous aspect, may be preserved in captivity for weeks, or even months, if properly supplied with food.

The *Ctenophora* swim in various positions, and some may often be noticed with their apical extremity turned downwards or forwards. Hence many writers term this the dorsal aspect; the digestive sac, by a strange perversion of language, being described as situate below the funnel; and so with the relative positions of the remaining organs. This practice is not only objectionable in itself, but has tended much to confuse almost every published account of the structure of a group of beings, than which few anatomical subjects are at once so easy and so accessible. Not in learning, but in unlearning, is the student of the *Ctenophora* compelled to waste his time and ingenuity.

Some *Ctenophora* are phosphorescent. In a species of *Bolina* common around our shores this beautiful property may very readily be observed. Specially distinguished for its luminosity is the much larger *Cestum Veneris* of the Mediterranean, which is said to gleam at night like a brilliant band of flame, moving beneath the surface of the water.

By Gegenbaur the *Ctenophora* have been divided into five families, which may be defined as follows:—

Order CTENOPHORA.

Sub-order 1. *Stenostomata*.

Family 1. CALYMMIDÆ.
Body furnished with a pair of antero-posterior oral lobes, and other smaller lateral appendages. *Tentacles* various, turned towards the mouth.

Family 2. CESTIDÆ.
Body ribbon-shaped, extended in a lateral direction, without oral lobes. *Tentacles* two in number, antero-posterior, turned towards the mouth.

Family 3. CALLIANIRIDÆ.
Body produced into a pair of wing-like, lateral lobes, bearing the ctenophores. *Tentacles* two in number, lateral, turned from the mouth.

Family 4. PLEUROBRACHIADÆ.
Body oval or spheroidal, without oral

lobes. *Tentacles* two in number, lateral, turned from the mouth.

Sub-order 2. *Eurystomata.*

Family 5. BEROIDÆ.
Body oval, elongated, without oral lobes. *Tentacles* absent.

Here we have slightly modified the definitions of Gegenbaur, at the same time indicating what appears to be the most natural sequence of the several families. The group *Callianiridæ* must for the present be considered as merely provisional. The four other divisions of *Ctenophora* have been recently elevated by Agassiz to the rank of sub-orders, and the entire number of families increased to ten. This arrangement, however, presents no advantage over the more simple and natural one adopted by Gegenbaur, which, in its turn, must be regarded as an improved modification of the prior classification of Eschscholtz.

SECTION IV.

DISTRIBUTION OF ACTINOZOA.

1. Relations to Physical Elements. — 2. Bathymetrical Distribution. — 3. Geographical Distribution.

1 **Relations to Physical Elements.** — All the *Actinozoa* are marine.

2. **Bathymetrical Distribution.** — Upon the whole it may be said that the *Alcyonaria* are less abundant between tide-marks, and occur in deeper

waters, than the *Zoantharia*. *Alcyonium* has been met with at seventy fathoms, but, like *Pennatula*, is common in much shallower seas. From so great a depth as 240 fathoms a species of *Virgularia*, *V. Finmarchica*, was dredged at Oxfjord by M. Sars, who also obtained, in the same locality, the widely different *Briareum grandiflorum*, a low creeping Alcyonid, allied to *Sarcodictyon*. The *Gorgonidæ*, in like manner, seem to prefer deep seas, *Corallium* having been found at 120, and *Gorgonia* itself at nearly 200 fathoms.

Though depths equal to or even exceeding those just mentioned have yielded many species of *Zoantharia*, *Ulocyathus*, for example, frequenting water of 200 fathoms, yet, in general, the members of this order are most abundant in seas of not more than 50 to 100 fathoms deep. The *Actinidæ* and *Madreporidæ* include those species which are most prone to descend below this limit. Many of the *Actinidæ*, it is well known, are numerous between tide-marks, the common Sea-anemone tending to encroach upon the line of high water.

The shallow vertical range of the reef-building *Actinozoa* has already been sufficiently explained. Certain species are chiefly restricted to particular parts of the reef; *Astræidæ* and *Seriatoporidæ* choosing its more submerged portions, below the outer exposed edge, upon which *Porites* and its allies flourish. On the surface of the reef both *Astræidæ* and *Fungidæ* may readily be distinguished, the labyrinthic form of *Mœandrina*, among other genera, being here especially conspicuous.

The soft-bodied non-adherent *Zoantharia* usually occur on muddy or sandy banks, at or near the level of low water. A few appear to be oceanic. *Philomedusa,* a minute form, from the Brazilian seas, habitually seeks shelter beneath the swimming-organ of various *Medusidæ* and *Lucernaridæ.*

The bathymetrical distribution of the *Ctenophora,* by reason of their oceanic habit, is scarcely amenable to observation. Some species, during the storms of winter, appear to seek considerable depths, on the return of spring again approaching to the surface.

3. **Geographical Distribution.** — The *Ctenophora, Alcyonaria,* and soft-bodied *Zoantharia* appear to be about equally abundant in tropical and temperate seas, many forms extending their range to high latitudes. Of coralligenous *Zoantharia* two families, *Turbinolidæ* and *Madreporidæ,* are not without northern and even arctic representatives, yet by far the majority of other sclerodermic species are seldom found to occur beyond the limits of the tropics. The reef-building Corals, according to Dana, will not flourish in water wherein the mean winter temperature is lower than 66° F. So that on either side of the equator a zone of water sufficiently heated for the growth of these Corals extends, the boundary lines of which have of necessity a somewhat contorted, irregular course, by reason of the varied combinations of circumstances influencing the local distribution of heat. Even within these limits other external conditions, not less essential than a high temperature to the welfare of reef-building Corals, are often absent. But when once the nature of these conditions has been carefully

understood, the many apparent anomalies in the distribution of Coral-reefs, far from being, as some have stated, unaccountable, become in each case susceptible of their appropriate physical explanation.

In the British seas about ten species of sclerodermic *Zoantharia* occur. The number of Mediterranean Corals is much greater, though these, with few exceptions, are specifically distinct from those observed by Ehrenberg in the Red Sea. The Mediterranean also yields two or three forms of sclerobasic *Zoantharia*, a group apparently unknown in more northern seas. *Corallium rubrum*, the Red Coral of commerce, would seem to be restricted to the same region, though other species of its genus have from time to time been dredged off Madeira and the Sandwich Isles.

Of *Actinozoa*, which occur beyond the limits of the Mediterranean and North Atlantic Seas, our knowledge still remains very imperfect, save only in the case of the reef-building Corals and the more conspicuous forms of *Ctenophora*. The genera *Cestum*, *Callianira*, *Calymma*, *Chiajea*, and *Leucothea* may be cited as examples of this order characteristic of the tropical and warmer temperate zones. *Ocyroe*, an obscure but interesting Ctenophorid, distinguished by the possession of two antero-posterior lobes, prolonged outwards at right angles to the true axis of the body, and which, when better known, may prove to be the immature condition of some apparently dissimilar form, has a range not wider than the equatorial regions of the Atlantic.

In high latitudes several *Actinidæ*, a few *Turbinolidæ* and *Madreporidæ*, together with various

Alcyonaria and *Ctenophora*, of which one genus, *Mertensia*, is said to be exclusively arctic, chiefly represent the class. The *Pennatulidæ* appear more numerous than other *Alcyonaria* around the northern colder temperate shores, seven species being named in the Norwegian fauna of Sars, while but three have yet been recorded as British. *Umbellularia*, a very aberrant member of this family, which presents a rod-like cœnosarc six feet in length, crowned with a spreading tuft of polypes at its summit, is only known from the published descriptions of two specimens, dredged from a depth of 236 fathoms, off the coast of Greenland, about the middle of the last century.

Among genera of *Actinozoa* which enjoy a wide distribution, *Actinia*, *Alcyonium*, *Zoanthus*, and *Gorgonia* are perhaps best worthy of mention. To this list should be added the names of several forms of *Ctenophora*, than which few marine animals appeared so well adapted to thrive under every variety of climatal conditions. Two genera in particular, *Beroe* and *Pleurobrachia*, are remarkable for their unbounded geographical area.

With less confidence can the names of such *Actinozoa* as are restricted in their range be, at present, insisted on. Renewed observations show that the number of extra-tropical genera, once thought to be peculiar to certain regions, must undergo considerable diminution. Of specific forms, however, not a few seem to characterise the various seas and shores to which they are confined.

The existence of natural barriers, whether of land or deep water, exercises a marked influence on the distribution of the *Actinozoa*. The differences between the East and West Indian species of Corals, or between the several Atlantic and Pacific forms of the class, often curiously resembling one another under similar conditions of depth and temperature, but, in a large number of cases, specifically distinct, may thus be easily accounted for. Many genera of fixed *Actinozoa*, abundant in one hemisphere, are found wholly wanting in the other. To a less extent is this observation true of the soft-bodied or free-swimming species.

Section V.

RELATIONS OF ACTINOZOA TO TIME.

1. General History of Actinozoa. — 2. History of Zoantharia. — 3. History of Rugosa. — 4. History of Alcyonaria. — 5, Silurian Corals. — 6. Devonian Corals. — 7. Carboniferous Corals. — 8. Permian Corals. — 9. Triassic Corals. — 10. Jurassic Corals. — 11. Cretaceous Corals. — 12. Tertiary Corals. — 13. Recent Actinozoa.

1. General History of Actinozoa. — *Actinozoa* appear to have been numerous during each of the greater artificial geologic epochs. The hard parts of the coralligenous species only have been preserved. Hence the expressions "fossil Corals" and "fossil Actinozoa" may be used as synonymous.

One order, *Ctenophora*, has no fossil representatives. The *Rugosa*, on the other hand, are wholly extinct.

The accompanying table exhibits, from a general point of view, the relations to time of the principal groups of *Actinozoa*. Lists are appended of those genera of Corals which range through more than one geological period.

ACTINOZOA.

Chronological Arrangement of Actinozoa.

Names of Groups.	Silurian.	Devonian.	Carboniferous.	Permian.	Triassic.	Jurassic.	Cretaceous.	Tertiary.	Recent.
Actinozoa	—	—	—	—	—	—	—	—	—
Zoantharia	—	—	—	—	—	—	—	—	—
Malacodermata									—
Sclerodermata	—	—	—	—	—	—	—	—	—
Aporosa	—				—	—	—	—	—
Turbinolidæ							—	—	—
Dasmidæ								—	
Oculinidæ							—	—	—
Stylophoridæ							—	—	—
Astræidæ						—	—	—	—
Echinoporidæ									—
Merulinidæ									—
Fungidæ	—					—	—	—	—
Perforata		—							
Madreporidæ							—	—	—
Poritidæ	—	—					—	—	—
Tubulosa	?	—	—						
Auloporidæ	?	—	—						
Tabulata	—	—	—	—	—	—	—	—	—
Thecidæ	—								
Seriatoporidæ									—
Favositidæ	—	—	—				—	—	—
Milleporidæ	—	—	—			—		—	—
Sclerobasica								—	—
Antipathidæ								—	—
Rugosa	—	—	—			—			
Cystiphyllidæ	—	—							
Cyathophyllidæ	—	—							
Cyathaxonidæ	—	—							
Stauridæ	—	—		—		—			
Alcyonaria	?						—	—	—
Alcyonidæ									—
Tubiporidæ									—
Pennatulidæ	?							—	—
Gorgonidæ	?							—	—
Ctenophora									—

PALEOZOIC CORALS,

WHICH OCCUR IN MORE THAN ONE GEOLOGICAL PERIOD.

Names of Genera arranged in order of their appearance.	Silurian.	Devonian.	Carboniferous.	Permian.
Cystiphyllum	—	—		
Syringophyllum	—	—		
Eridophyllum	—	—		
Ptychophyllum	—	—		
Aulacophyllum	—	—		
Heliolites	—	—		
Psammopora	—	—		
Aulopora	?	—		
Cyathaxonia	—		—	
Clisiophyllum	—	—	—	
Propora	—		—	
Cyathophyllum	—	—	—	
Zaphrentis	—	—	—	
Syringopora	—	—	—	
Favosites	—	—	—	
Emmonsia	—	—	—	
Alveolites	—	—	—	
Chætetes	—	—	—	—
Philippsastrea		—	—	
Lithostrotion		—	—	
Campophyllum		—	—	
Amplexus		—	—	
Lophophyllum		—	—	
Beaumontia		—	—	
Michelinia		—	—	
Fistulipora			—	—

MESOZOIC, CAINOZOIC, AND RECENT CORALS,
WHICH OCCUR IN MORE THAN ONE GEOLOGICAL PERIOD.

Names of Genera arranged in order of their appearance.	Names of Periods.				
	Triassic.	Jurassic.	Cretaceous.	Tertiary.	Recent.
Hybocœnia	—	—			
Goniocora	—	—			
Rhabdophyllia	—	—	—		
Cladophyllia	—	—	—		
Isastrea	—	—	—		
Montlivaultia	—	—	—	—	
Eurymeandra	—	—	—	—	
Thamnastrea	—	—	—	—	
Enallohælia		—	—		
Cœlosmilia		—	—		—
Pachygyra		—	—		
Rhipidogyra		—	—		
Stylosmilia		—	—		
Stylina		—	—		
Cyathophora		—	—		
Calamophyllia		—	—		
Haplophyllia		—	—		
Adelastræa		—	—		
Pleurocœnia		—	—		
Trochocyathus		—	—	—	
Trochosmilia		—	—	—	
Astrocœnia		—	—	—	
Stephanocœnia		—	—	—	
Thecosmilia		—	—	—	
Oroseris		—	—	—	
Mæandrina		—	—	—	—
Favia		—	—	—	—
Heliastræa		—	—	—	—
Ulophyllia		—		—	—
Plerastræa		—		—	—
Millepora		—			—

MESOZOIC, CAINOZOIC, AND RECENT CORALS, WHICH OCCUR IN MORE THAN ONE GEOLOGICAL PERIOD — *continued*.

Names of Genera arranged in order of their appearance.	Triassic.	Jurassic.	Cretaceous.	Tertiary.	Recent.
Barysmilia			—	—	
Stylocœnia			—	—	
Centrocœnia			—	—	
Phyllocœnia			—	—	
Brachyphyllia			—	—	
Rhizangia			—	—	
Cyclolithus			—	—	
Stephanophyllia			—	—	
Bathycyathus			—		—
Lophosmilia			—		—
Diploria			—		—
Leptoria			—		—
Goniastræa			—		—
Cyphastræa			—		—
Pavonaria			—		—
Caryophyllia			—	—	—
Mycetophyllia			—	—	—
Hydnophora			—	—	—
Cladocora			—	—	—
Cycloseris			—	—	—
Corallium			—	—	—
Isis			—		
Acanthocyathus				—	—
Paracyathus				—	—
Sphenotrochus				—	—
Desmophyllum				—	—
Oculina				—	—
Lophohælia				—	—
Stylophora				—	—
Euphyllia				—	—
Galaxea				—	—
Lithophyllia				—	—
Dasyphyllia				—	—

MESOZOIC, CAINOZOIC, AND RECENT CORALS, WHICH OCCUR IN MORE THAN ONE GEOLOGICAL PERIOD — *continued*.

Names of Genera arranged in order of their appearance.	Triassic.	Jurassic.	Cretaceous.	Tertiary.	Recent.
Symphyllia				—	—
Plesiastræa				—	—
Solenastræa				—	—
Astræa				—	—
Prionastræa				—	—
Astrangia				—	—
Phyllangia				—	—
Eupsammia				—	—
Endopachys				—	—
Balanophyllia				—	—
Dendrophyllia				—	—
Madrepora				—	—
Turbinaria				—	—
Astræopora				—	—
Porites				—	—
Rhodaræa				—	—
Virgularia				—	—
Mopsea				—	—
Hyalopathes				—	—

Chœtetes passes up into the Trias. With this exception no genus of Corals survives the Paleozoic epoch except, perhaps, *Isis*, of which doubtful indications have been met with in rocks of very ancient date.

No Triassic genus of Corals has recent representatives. Of genera which occur in the Jurassic series seven still survive. Fourteen recent genera first appear in the Chalk, while very many are common to the Tertiary and Recent periods. But few Recent species of Corals occur in a fossil state.

It appears also from the preceding tables that six genera of Corals range through four periods, thirty-four through three, and sixty-eight through two. Some genera, however, arise in one formation, are apparently absent from the next, but again present themselves at a subsequent period. Of this seeming anomaly *Millepora* furnishes an example. Such instances must always be received with suspicion, since they are probably due to defective observation.

2. **History of Zoantharia.**—

All extinct *Zoantharia* belong to the group of *Sclerodermata*, with the exception of a few slight indications of *Antipathidæ* which appear in the Tertiary period. The *Malacodermata* are wholly recent. On the other hand, the small group of *Tubulosa* does not survive the Paleozoic epoch. But two families of *Zoantharia*, *Thecidæ* and *Auloporidæ*, have altogether disappeared. On the whole it may be said that *Tabulata* prevail in the Paleozoic deposits, *Aporosa* and *Perforata* in those which succeed. *Tabulata* are comparatively scarce in strata anterior to the Carboniferous, though no geological period is without some representative of this division, and in modern seas four genera have been observed. A single genus, *Palæocyclus*, which occurs in the Silurian period, is the only known representative of *Aporosa* in strata older than the Trias. The *Perforata* are represented in the Paleozoic rocks by two genera, but, excepting these, no other forms of the group have been met with in deposits of earlier date than the Jurassic.

3. **History of Rugosa.—**

All *Rugosa* are confined to the Paleozoic epoch, with the exception of the genus *Holocystis*, which is peculiar to the Lower Greensand, where it is represented by a single species, *H. elegans*.

The *Rugosa* first appear in the Lower Silurian. They are especially abundant in the Upper Silurian, Devonian, and Carboniferous deposits. In the Permian rocks they are represented by only one generic form.

4. **History of Alcyonaria.—**

Few genera of *Alcyonaria* have hitherto been found in a fossil condition, and scarcely three of these are wholly extinct. The existence of this order during the Paleozoic epoch must be regarded as doubtful, though the genus *Protovirgularia* has been constituted for the reception of a Silurian fossil, supposed to belong to the family *Pennatulidæ*. The genus *Isis*, also, has been stated to occur in some of the Paleozoic formations. With these exceptions no *Alcyonaria* have been found in rocks more ancient than the Chalk. The family *Alcyonidæ* is without an extinct representative.

5. **Silurian Corals.—**

The Silurian Corals consist chiefly of *Rugosa* and *Tabulata*. The *Aporosa* are represented by the genus *Palæocyclus;* the *Perforata* by *Protaræa*. Doubtful indications of *Tubulosa* and *Alcyonaria* also occur. At least nine families of Corals first make their appearance in this period, and one,

Thecidæ, does not survive it. The following genera are peculiar to the Silurian series:

FUNGIDÆ.
  Palæocyclus.
PORITIDÆ.
  Protaræa.
THECIDÆ.
  Thecia.
  Columnaria.
MILLEPORIDÆ.
  Lyellia.
CYATHOPHYLLIDÆ.
  Streptelasma.
  Omphyma.

Goniophyllum.
Strombodes.
FAVOSITIDÆ.
  Cœnites.
  Halysites.
  Fletscheria.
  Danaia.
  Dekaia.
  Labecheia.
  Constellaria.
STAURIDAE.
  Stauria.

6. Devonian Corals.—

Excepting the genus *Aulopora*, and the ambiguous form *Pleurodictyum*, the Devonian Corals consist wholly of *Rugosa* and *Tabulata*. One family, *Seriatoporidæ*, first makes its appearance in this formation, and one, *Cystiphyllidæ*, does not survive it. The following genera are exclusively Devonian:

PORITIDÆ.
  Pleurodictyum.
AULOPORIDÆ.
  Aulopora.
SERIATOPORIDÆ.
  Dendropora.
  Trachypora.
FAVOSITIDÆ.
  Thecostegites.
  Chonostegites.
  Rœmeria.
MILLEPORIDÆ.
  Battersbyia.

CYATHOPHYLLIDÆ.
  Smithia.
  Spongophyllum.
  Acervularia.
  Endophyllum.
  Pachyphyllum.
  Heliophyllum.
  Chonophyllum.
  Anisophyllum.
  Baryphyllum.
  Hadrophyllum.
  Hallia.
  Combophyllum.
STAURIDÆ.
  Metriophyllum.

7. Carboniferous Corals.—

In addition to the genus *Pyrgia*, the Coral fauna of the Carboniferous rocks seems to be wholly made up of *Rugosa* and *Tabulata*. Three families, *Auloporidæ*, *Cyathophyllidæ*, and *Cyathaxonidæ*, do not outlive this period. The following genera are restricted to the Carboniferous deposits:

AULOPORIDÆ.	CYATHOPHYLLIDÆ:
Pyrgia.	*Lonsdaleia*.
SERIATOPORIDÆ.	*Stylaxis*.
Rhabdopora.	*Chonaxis*.
CYATHOPHYLLIDÆ.	*Aulophyllum*.
Axophyllum.	*Menophyllum*.
	Trochophyllum.

8. **Permian Corals.**—

The few Permian Corals hitherto found belong to the *Rugosa* and *Tabulata*. The genus *Polycœlia*, of the family *Stauridæ*, is peculiar to this period.

9. **Triassic Corals.**—

Fossil remains of Corals are scarce in the Trias. The family *Astrœidæ*, so abundantly represented in all subsequent formations, now first makes its appearance. To this group most of the Triassic Corals have been referred. The *Favositidæ* are represented by the old genus *Chœtetes*. It can scarcely be said that any genera of Corals are characteristic of this formation.

10. **Jurassic Corals.**—

There are no *Rugosa* in Jurassic rocks, and *Millepora*, a recent genus, is the sole representative of the *Tabulata*. The greater number of Jurassic Corals belong to the *Aporosa*, and certain beds of this series have received the name of Coral-Rag from the great abundance of *Astrœidæ* which they contain. The genera *Stylina* and

Montlivaultia are especially rich in species. Two generic forms represent the *Perforata*. The families *Turbinolidæ* and *Oculinidæ* now appear for the first time. The following genera are exclusively Jurassic:

TURBINOLIDÆ.
  Discocyathus.
  Thecocyathus.
OCULINIDÆ.
  Eulalia.
ASTRÆIDÆ.
  Axosmilia.
  Haplosmilia.
  Phytogyra.

ASTRÆIDÆ:
  Placosphyllia.
  Angeastræa.
FUNGIDÆ.
  Protoseris.
  Comoseris.
PORITIDÆ.
  Microsolena.
  Anomophyllum.

11. **Cretaceous Corals.**—

The Corals of the Chalk are very numerous, belonging chiefly to the *Aporosa* and *Perforata*. Here also undoubted indications of *Alcyonaria* present themselves. The *Tabulata* are represented by two genera. For the last time the order *Rugosa* makes its appearance, a single genus, *Holocystis*, being its representative. The families *Madreporidæ, Pennatulidæ* (?), and *Gorgonidæ* (?) now first appear. The following genera are peculiar to this period:

TURBINOLIDÆ.
  Brachycyathus.
  Cyclocyathus.
  Stylocyathus.
  Smilotrochus.
OCULINIDÆ.
  Synhelia.
  Baryhelia.
ASTRÆIDÆ.
  Placosmilia.
  Diploctenium.
  Parasmilia.
  Peplosmilia.

ASTRÆIDÆ:
  Holocœnia.
  Acanthocœnia.
  Placocœnia.
  Elasmocœnia.
  Pentacœnia.
  Heterocœnia.
  Leptophyllia.
  Dactylosmilia.
  Hymenophyllia.
  Aspidiscus.
  Stelloria.
  Meandrastræa.

ASTRÆIDÆ:
   Dimorphastræa.
   Pleurocora.
FUNGIDÆ.
   Micrabacia.
   Anabacia.
   Genabacia.

MADREPORIDÆ.
   Actinacis.
FAVOSITIDÆ.
   Koninckia.
MILLEPORIDÆ.
   Polytremacis.
STAURIDÆ.
   Holocystis.

12. **Tertiary Corals.**—

The Tertiary formations are abundantly supplied with Corals, chiefly belonging to the *Aporosa* and *Perforata*. The *Tabulata* are represented by a single genus. There are distinct traces of *Alcyonaria*. The Sclerobasic *Zoantharia* now first present themselves. Here, too, appear for the first time the *Dasmidæ* and *Stylophoridæ*, whose claim to the rank of distinct families is somewhat doubtful. The *Dasmidæ* do not survive the period. The following genera are restricted to the Tertiary deposits:

TURBINOLIDÆ.
   Conocyathus.
   Deltocyathus.
   Leptocyathus.
   Ecmesus.
   Turbinolia?
   Platytrochus.
   Ceratotrochus.
   Discotrochus.
DASMIDÆ.
   Dasmia.
OCULINIDÆ.
   Diplohelia.
   Astrohelia.
STYLOPHORIDÆ.
   Arœacis.
ASTRÆIDÆ.
   Cyclosmilia.
   Dendrosmilia.

   Haplocœnia.
   Circophyllia.
   Tichastræa.
   Metastræa.
   Cryptangia.
   Cladangia.
FUNGIDÆ.
   Trochoseris.
   Cyathoseris.
MADREPORIDÆ.
   Lobopsamnia.
   Stereopsamnia.
   Dendracis.
PORITIDÆ.
   Litharœa.
MILLEPORIDÆ.
   Axopora.
PENNATULIDÆ.
   Graphularia.

13. **Recent Actinozoa.**—
Except the *Rugosa, Tubulosa* and *Thecidæ,* all the orders and families of the sub-kingdom *Cœlenterata* have living representatives. The names of the recent genera are too numerous to be here mentioned. Many of them have already been indicated in those parts of the work devoted to the study of their classification.

The systematic form under which we have sought to exhibit the above selection of facts touching the general relations to time of the several groups of Corals must not lead the student to repose too much trust in a record confessedly so imperfect, or regard it as aught else than, in the strictest sense, provisional. In particular it would appear from certain investigations, not yet fully published, that the supposed line of demarcation between the Paleozoic and Neozoic Coral forms does not really exist. The entire subject, like many others discussed in the preceding pages, still offers a wide and richly promising field for future inquiry.

BIBLIOGRAPHY OF THE CŒLENTERATA.

(1.) FREY und LEUCKART.—'Beiträge zur Kenntniss Wirbelloser Thiere,' 1847. (pp. 1—40.)
(2.) LEUCKART.—'Ueber die Morphologie der Wirbellosen Thiere,' 1848. (pp. 13—31.)
(3.) LEYDIG.—'Lehrbuch der Histologie,' 1857. (passim.)
(4.) CARUS, J. V.—'Icones Zoötomicæ,' 1857. (Taf. II.—IV.)
(5.) GEGENBAUR.—'Grundzüge der Vergleichenden Anatomie,' 1859. (Zweiter Abschnitt, pp. 67—103.)
(6.) BRONN.—'Die Klassen und Ordnungen des Thier-reichs,' 1859-60. (Zweiter Band, Lief. 1—6.)
Also, the systematic works of CUVIER, Règne Animal (ed. with plates, by his pupils); LAMARCK, Hist. Nat. des Animaux s. Vertèbres (2nd ed. by Deshayes and Milne Edwards); DE BLAINVILLE, Manuel d'Actinologie, 1834; JOHNSTON, History of British Zoöphytes, 1847, 2nd ed.; CAVOLINI, Memorie per servire alla storia de' Polipi marini, 1785; Bosc, Hist. Nat. des Vers, 1802; ESPER, Die Pflanzen-Thiere, 1806; PALLAS, Elenchus Zoöphytorum, 1766, Charakteristik der Thierpflanzen, 1787; and the following, among other, miscellaneous treatises:
(7.) BOHADSCH.—'De quibusdam Animalibus marinis,' 1761.
(8.) BUSCH.—'Beobachtungen über Anatomie und Entwickelung einiger Wirbellosen Seethiere,' 1851.
(9.) DALYELL.—'Rare and Remarkable Animals of Scotland,' 1847-8.
(10.) DELLE CHIAJE.—'Descrizione e Notomia degli Animali invertebrati della Sicilia citeriore osservati vivi negli anni 1822-30,' 1841-44.
(11.) FORBES and GOODSIR.—'On some remarkable Marine Invertebrata new to the British Seas,' Trans. Roy. Soc. Edin. 1851.

(12.) FORSKAL.—'Descriptiones Animalium,' 1775.
(13.) ——————.—'Icones Rerum naturalium,' 1776.
(14.) GOSSE.—'A Naturalist's Rambles on the Devonshire Coast,' 1853.
(15.) LAURENT.—'Zoöphytologie' (in Vaillant's Voyage de la Bonite), 1844.
(16.) LESSON.—'Centurie Zoölogique,' 1830.
(17.) LEUCKART.—'Ueber den Polymorphismus der Individuen,' 1851.
(18.) MÜLLER, O. F.—'Zoölogia Danica,' 1788-1806.
(19.) QUOY et GAIMARD.—'Zoölogie' (in Voyage de l'Uranie, sous Freycinet), 1824.
(20.) ——————————.—'Zoölogie' (in Voyage de l'Astrolabe, sous Dumont d'Urville), 1830-4.
(21.) SARS.—'Beskrivelser og Jagttagelser, &c.' 1835.
(22.) ——.—'Bidrag til Kundskaben om Middlehavets Littoral-Fauna,' 1857.
(23.) ——, KOREN et DANIELSSEN.—'Fauna littoralis Norvegiæ,' 1846 and 1856.
(24.) STEENSTRUP.—'On the Alternation of Generations' (Eng. Trans. by Busk), 1845.
(25.) STIMPSON.—'Synopsis of the Marine Invertebrata of Grand Manan' (sep. and in Smith. Cont.), 1853.

It is unnecessary any longer to prolong the above list, since a complete enumeration of the various memoirs on Cœlenterata may be found in the 'Bibliotheca Zoologica' of Carus and Engelmann (Leipzig, 1861), pp. 320—45. In the 'Natural History Review' (London, 1861 et seq.), continuations of this catalogue will from time to time appear. The Reports furnished each year by Leuckart to Wiegmann's Archiv. für Naturgeschichte may also be consulted with advantage. Of the more select memoirs which treat of particular groups of Cœlenterata we have here attempted to subjoin the names:

HYDROZOA.

a. HYDRIDÆ.

(26.) TREMBLEY.—'Mémoires pour servir à l'histoire d'un genre de Polypes d'eau douce, à bras en forme de cornes,' 1744.
(27.) HANCOCK.—'Notes on a species of Hydra found in the Northumberland Lakes,' A. N. H. 1850.

(28.) THOMSON, ALLEN.—'On the Co-existence of Ovigerous and Spermatic Capsules on the same individuals of the Hydra viridis,' Phil. Journ. 1847.

(29.) ECKER.—'Zur Lehre von Bau und Leben der contractilen Substanz der niedersten Thiere,' Z. W. Z. 1849 (or Trans. by Busk in Q. J. M. S. 1854).

(30.) JÄGER.—' Ueber das spontane Zerfallen der Süsswasserpolypen nebst einigen Bemerkungen über Generationswechsel,' Vien. Sitz. 1860: and other memoirs cited in Bib. Zoöl., especially those of ALLMAN, CORDA, LAURENT, LEYDIG, and ROUGET. The British species of Hydra are described by JOHNSTON (op. s. cit.), and LEWES, A. N. H. 1860.

b. CORYNIDÆ AND SERTULARIDÆ.

(31.) LOVÉN.—' Beitrag zur Kenntniss der Gattungen Campanularia und Syncoryne,' Wiegm. Arch. 1837. (Abstract in STEENSTRUP) (24).

(32.) BENEDEN, VAN.—' Mémoire sur les Campanulaires de la côte d'Ostende, considérés sous le rapport physiologique, embryogénique et zoölogique.' Bruss. Mém. 1843.

(33.) ————, ——.—' Recherches sur l'embryogénie des Tubulaires et l'histoire naturelle des différents genres de cette famille qui habitent la côte d'Ostende,' Bruss. Mém. 1844.

(34.) SCHULTZE, MAX.—' Ueber die männlichen Geschlechtstheile der Campanularia geniculata,' Arch. Anat. 1850 (or Trans. in Q. J. M. S. 1855).

(35.) MUMMERY.—'On the Development of Tubularia indivisa,' Q. J. M. S. 1853.

(36.) ALLMAN.—'On the Anatomy and Physiology of Cordylophora,' Phil. Trans. 1853.

(37.) ————.—' On the Structure of the Reproductive Organs in certain Hydroid Polypes,' R. S. E. Proc. 1857-8; and ' Additional Observations on the Morphology of the Reproductive Organs in the Hydroid Polypes,' ibid. 1858.

(38.) ————.—' Notes on the Hydroid Zoophytes,' A. N. H. 1859 et seq.

Other memoirs on the reproductive organs and development of these orders are those of DUJARDIN, Ann. S. N. 1843 and 45; DESOR, Ann. S. N. 1849; FORBES, A. N. H. 1844; LISTER, Phil. Trans. 1834; KÖLLIKER, Z. W. Z. 1853; and KROHN, Arch. Anat. 1843 and 53, Wiegm. Arch. 1851. Also, the works of SARS (21), (22), (23); STEENSTRUP (24); and GEGENBAUR (47). No complete monograph of the genera of Corynidæ and Sertularidæ has yet appeared. For descriptions and figures of the British species see the works of DALYELL (9); GOSSE (14), and Linn. Trans. 1857; ELLIS, 'Essay towards a Natural History of Corallines,' 1755, and JOHNSTON (op. s. cit.); together with the papers of ALLMAN; ALDER, A. N. H. 1856 et seq.; HINCKS, A. N. H. 1851 et seq.; STRETHILL WRIGHT, Phil. Journ. 1857-8-9; and WYVILLE THOMPSON, A. N. H. 1853. American forms have been described by AYRES, Bost. N. H. S. 1852 and 5; DESOR, Bost. N. H. S. 1848-9; M'CRADY (49); MURRAY, A. N. H. 1860; and STIMPSON (25): Southern species by BUSK, B. Ass. Rep. 1850, in Q. J. M. S. passim, and in supplement to Vol. I. of Voyage of 'Rattlesnake;' and HINCKS, A. N. H. 1861.

c. CALYCOPHORIDÆ AND PHYSOPHORIDÆ.

(39.) HUXLEY.—'The Oceanic Hydrozoa—A Description of the Calycophoridæ and Physophoridæ observed during the voyage of H. M. S. "Rattlesnake" in the years 1846-50. With a General Introduction,' 1859: and the writings of MILNE EDWARDS, GEGENBAUR, KÖLLIKER, LEUCKART and VOGT, cited in the bibliography appended to the same work. Other memoirs by GEGENBAUR, Nov. Act. 1860; CLAUS, Z. W. Z. 1860; and KEFERSTEIN und EHLERS (Abstract in Wiegm. Arch. 1860), have since appeared.

d. MEDUSIDÆ AND LUCERNARIDÆ.

(40.) ESCHSCHOLTZ.—'System der Acalephen,' 1829.
(41.) EHRENBERG.—'Die Akalephen des rothen Meeres und der Organismus der Medusen der Ostsee,' 1836.
(42.) WAGNER.—'Ueber den Bau der Pelagia noctiluca, und die Organisation der Medusen,' 1841.
(43.) LESSON.—'Acalephes,' Nouvelles Suites à Buffon, 1843.
(44.) FORBES.—'A Monograph of the British Naked-eyed Medusæ,' 1848.
(45.) AGASSIZ.—'On the Naked-eyed Medusæ of the Shores of Massachusetts, in their Perfect State of Development,' Trans. Amer. Acad. 1849.
(46.) HUXLEY.—'On the Anatomy and Affinities of the Family of the Medusæ,' Phil. Trans. 1849.

(47.) GEGENBAUR.—'Zur Lehre vom Generationswechsel und der Fortpflanzung bei Medusen und Polypen,' 1854.

(48.) ——————.—'Versuch eines Systemes der Medusen, mit Beschreibung neuer oder wenig gekannter Formen,' Z. W. Z. 1857.

(49.) M'CRADY.—'Gymnophthalmata of Charleston Harbor,' Ell. Soc. Proc. 1857: and the systematic papers of PERON et LESUEUR, Ann. d. Mus. 1809-10; BRANDT, Petersb. Mem. 1833 and 38; and LUTKEN, Videns. Med. 1850; with other memoirs by ALLMAN (Structure of Lucernariadæ), Q. J. M. S. 1860; MILNE EDWARDS (Structure of Æquorea) (67), and (Circulation in Lucernaridæ), A. S. N. 1845; DERBÉS (Hermaphroditism of Chrysaora), Ann. S. N. 1850; KÖLLIKER (Medusidæ of Messina), Z. W. Z. 1853; LEUCKART (Medusidæ of Nice), Wiegm. Arch. 1856; FRITZ MÜLLER (Gastric filaments of Lucernaridæ), Z. W. Z. 1858; EYSENHARDT (Rhizostoma) Nov. Act. 1821; and TILESIUS (Cassiopeia), Nov. Act. 1831.

On Structure of Marginal Bodies see especially GEGENBAUR, Arch. Anat. 1856 (or English abstract in Q. J. M. S. 1858).

On Minute Structure of Medusidæ and Lucernaridæ, vid. BUSK, Mic. Trans. 1852; SCHULTZE, Arch. Anat. 1856; and HUXLEY (46).

On Structure of Charybdeidæ: MILNE EDWARDS (Charybdea), Ann. S. N. 1833; and FRITZ MÜLLER (Tamoya), Halle Abh. 1859.

On Development of Medusidæ: J. MÜLLER (Æginopsis), Arch. Anat. 1851; GEGENBAUR (Cunina and Trachynema), (47); FRITZ MÜLLER (Liriope), Wiegm. Arch. 1859; M'CRADY (Cunina), Ell. Soc. Proc. 1856; and CLAPARÈDE, Z. W. Z. 1860.

On Development of Lucernaridæ: SARS (21), Isis, 1833, and Wiegm. Arch. 1837-41 and 1857; SIEBOLD, Beitrage zur Naturgeschichte der Wirbellosen Thiere, 1839; STEENSTRUP (24); DALYELL (9); DESOR, Ann. S. N. 1849; and REID, 'Physiological, Pathological, and Anatomical Researches,' or A. N. H. 1848. These writers treat chiefly of Aurelia, Cyanea, and Chrysaora. For the development of other genera see GEGENBAUR (Cassiopeia), (47); FRANTZIUS (Cephea), Z. W. Z. 1853; and KROHN (Pelagia), Arch. Anat. 1855 (or English abstract in A. N. H. 1856).

The British species of Medusidæ (and Medusoids) are illustrated by FORBES (44), and Zoöl. Proc. 1851; FORBES and GOODSIR (11); GOSSE (14); GREENE, Nat. Hist. Rev. 1857-8; COBBOLD, Q. J. M. S. 1858; PATTERSON, D. U. Z. B. A. 1859; and STRETHILL WRIGHT, Phil. Journ. 1859. The Lucernariadæ are described by JOHNSTON (op. s. cit.); OWEN, B. Ass. Rep. 1849; GOSSE, A. N. H. 1860; and ALLMAN, op. s. cit. and A. N. H. 1860. Brief descriptions, without figures, of the pelagic Lucernaridæ are given by FORBES (44), but most of the species are figured in the other works mentioned above.

ACTINOZOA.

a. ZOANTHARIA.

(50.) DICQUEMARE.—'Essay towards elucidating the history of the Sea-Anemones,' Phil. Trans. 1773. 'A second essay on the natural history of the Sea-Anemones,' ibid. 1775, and a third essay in ditto, 1777.

(51.) ELLIS and SOLANDER.—'The Natural History of many curious and uncommon Zoöphytes,' 1786.

(52.) RAPP.—'Ueber die Polypen im Allgemeinen und die Actinien insbesondere,' 1829.

(53.) EHRENBERG.—'Beitrage zur physiologischen Kenntniss der Corallenthiere im Allgemeinen, und besonders des rothen Meeres,' and 'Ueber die Natur und Bildung der Coralleninseln und Corallenbänke im rothen Meere,' Berl. Abh. 1834.

(54) QUATREFAGES.—'Mémoire sur les Edwardsies,' Ann. S. N. 1842.

(55.) DANA.—'Report on Zoöphytes,' and 'Atlas of Zoöphytes,' (U. S. Exploring Expedition), 1849. The introductory part of this Report (which is now out of print) has been published under the title of "Structure and Classification of Zoöphytes." A "Synopsis" of the Report itself has since appeared.

(56.) EDWARDS et HAIME.—'Recherches sur les Polypiers,' Ann. S. N. 1848—52.

(57.) ————.—'Histoire Naturelle des Coralliaires ou Polypes proprement dits,' 1857—60.

(58.) HOLLARD.—'Monographie anatomique du genre Actinia de Linné, considéré comme type du groupe général des Polypes Zoanthaires,' Ann. S. N. 1851.

(59.) HAIME.—'Mémoire sur le Cérianthe,' Ann. S. N. 1854.
(60.) GOSSE.—'Actinologia Britannica: A History of the British Sea-anemones and Madrepores,' 1860.

See also various memoirs by MILNE EDWARDS, HOLLARD, and GOSSE, cited in Bib. Zoöl., in addition to those of SPIX; KÖLLIKER; LEWES, 'Sea-side Studies;' LACAZE DU THIERS; RATHKE; ERDL; and others.

On the structure of Actinia see especially HOLLARD (58); GOSSE (60); HAIME, C. rend. 1854; FREY und LEUCKART (1); TEALE, Trans. Leeds Soc. 1837, and B. Ass. Rep. 1838; and COBBOLD, A. N. H. 1853.

On the Sclerobasic Zoantharia, vid. BRANDT,' Symbolæ ad Polypos Hyalochætides spectantes, 1859;' and SCHULTZE, C. rend. 1860. DANA (55) describes the polypes of Antipathes.

(57) is a complete monograph of the orders Zoantharia, Alcyonaria, and Rugosa. For collections of figures the works of ESPER, ELLIS, DANA, and the French voyagers must be chiefly consulted. The British Zoantharia are described and figured by GOSSE (60).

On Coral reefs and islands see DARWIN, 'The Structure and Distribution of Coral Reefs,' 1842 (now forming part of the same author's "Geological Observations"); and DANA, 'On Coral Reefs and Islands,' 1853.

b. ALCYONARIA.

(61.) RAPP.—'Untersuchungen über den Bau einiger Polypen des Mittelandischen Meeres,' Nov. Act. 1829.
(62.) EDWARDS, MILNE.—'Mémoire sur un nouveau genre de la famille des Alcyoniens,' and 'Observations sur les Alcyons proprement dits,' in Recherches sur les Polypes, 1838, or Ann. S. N. 1838.
(63.) AGASSIZ.—'On the Structure of the Halcyonoid Polypes,' Am. Ass. Rep. 1850. Also: MILNE EDWARDS, in Règne Animal, and several of the works which treat of Zoantharia. JOHNSTON, op. s. cit., describes and figures the British Alcyonaria.

c. RUGOSA and FOSSIL CORALS.

(64.) EDWARDS et HAIME.—'A Monograph of the British Fossil Corals' (published by the Palæontographical Society), 1850 —55.

(65.) EDWARDS et HAIME.—'Monographie dês Polypiers Fossiles des Terrains Paléozoïques,' Arch. d. Mus. 1851.

And many other palæontological works, most of which are quoted by EDWARDS et HAIME (57).

d. CTENOPHORA.

(66.) MERTENS.—'Beobachtungen über die Beroëartigen Acalephen,' Petersb. Mem. 1833.

(67.) EDWARDS, MILNE.—'Observations sur divers Acalèphes,' Ann. S.N. 1841.

(68.) ——————————.—'Note sur l'appareil Gastrovasculaire de quelques Acalèphes Cténophores, Ann. S. N. 1857.

(69.) WILL.—'Horæ Tergestinæ, oder Beschreibung und Anatomie der im Herbste 1843, bei Triest beobachteten Akalephen.' 1844.

(70.) AGASSIZ.—'On the Beroid Medusæ of the Shores of Massachusetts, in their Perfect State of Development,' Amer. Acad. Trans. 1849.

(71.) ——————.—'Contributions to the Natural History of the United States of America,'—Part II. of Second Monograph, 1860.

(72.) GEGENBAUR.—'Studien über Organisation und Systematik der Ctenophoren,' Wiegm. Arch. 1856: and the papers of GRANT, Z. Trans. 1833; KÖLLIKER, Z. W. Z. 1853; LESSON, Ann. S. N. 1836; QUOY et GAIMARD, Ann. S. N. 1825; RANG, Bourd. Soc. Lin. 1826, and Isis, 1832: and WAGNER, Arch. Anat. 1847. The systematic works of ESCHSCHOLTZ (40); DE BLAINVILLE, op. s. cit.; and LESSON (43) may also be consulted. On Development of Ctenophora see the papers of M'CRADY, (Beroe and Bolina), Ell. Soc. Proc. 1859; PRICE (Pleurobrachia), B. Ass. Rep. 1846; SEMPER (Eucharis), Z. W. Z. 1858; and STRETHILL WRIGHT (Pleurobrachia), Phil. Journ. 1856. The British Ctenophora are described by FORBES and GOODSIR, B. Ass. Rep. 1840-1; and PATTERSON, R. I. A. Trans. 1839-40.

. The abbreviations above used to indicate the titles of periodical journals are explained in the Bibliography of 'Natural History Review,' 1861.

QUESTIONS ON THE CŒLENTERATA.

1. By what structural features are Cœlenterata separated from other primary divisions of the animal kingdom?
2. Contrast the two sub-kingdoms, Protozoa and Cœlenterata.
3. Describe the typical structure of a thread-cell.
4. Compare the classes, Hydrozoa and Actinozoa.
5. Describe the structure of Hydra. How does the 'polypite' of this genus differ from that of the non-budding forms of the Corynidæ?
6. Define the terms,
    a. 'hydrotheca';
    b. 'hydrophyllium';
    c. 'hydrocyst.'
7. In what order of Hydrozoa do 'nematophores' occur? Describe the structure and position of these appendages.
8. Describe the structure of a 'nectocalyx.' How does this organ differ from an 'umbrella'?
9. Describe the structure and relations of the nectocalyces in Diphyes, and state how the same parts are modified in Praya.
10. In what Physophoridæ are nectocalyces absent?
11. Briefly describe the modifications of the cœnosarc, and relative attachment of its appendages, in the following genera of Physophoridæ:
    a. Physophora;
    b. Stephanomia;
    c. Apolemia;
    d. Athorybia;
    e. Velella.

12. Compare the structure of the tentacles in
    a. Physalia;
    b. Forskalia;
    c. Apolemia.
13. What is the position of the tentacles in the following genera of Hydrozoa:
    a. Tubularia;
    b. Hydractinia;
    c. Diphyes;
    d. Porpita;
    e. Pelagia?
14. Describe the 'pneumatocyst' of any of the Physophoridæ, and the principal modifications which it presents among other genera of the same order.
15. Describe, as to structure and position,
    a. the marginal 'vesicle' of Geryonia;
    b. the 'lithocyst' of a free Lucernarid.
16. How are the 'gonophores' situated in the following genera of Corynidæ:
    a. Tubularia;
    b. Hydractinia;
    c. Clavatella;
    d. Cordylophora?
17. Compare the structural relations of the reproductive organs in
    a. Lucernaria;
    b. Aurelia;
    c. Rhizostoma.
18. What are 'gonoblastidia'? Explain the modifications which these structures present among the Sertularidæ.
19. In what genera of Lucernaridæ do free reproductive zoöids occur? Trace the development of any of these forms.
20. Describe the development of a medusiform gonophore.
21. Give some account of the early stages of development in
    a. Cordylophora lacustris;
    b. Campanularia Loveni;
    c. Cunina octonaria.
22. What rule seems to govern the successive development of the appendages among the Physophoridæ?
23. Define the orders:
    a. Sertularidæ;
    b. Calycophoridæ;
    c. Lucernaridæ.

24. What genera of Cœlenterata are known to inhabit fresh water?
25. Give some account of the geographical distribution of the Sertularidæ.
26. Describe the minute structure of the body-wall in Actinia.
27. What law appears to determine the number of parts among the several orders of Actinozoa?
28. What is the number and structure of the tentacles in the Alcyonaria?
29. Describe the structure of a typical 'corallite.'
30. Define the order Rugosa.
31. Explain the formation of the gyrate corallum of Mœandrina.
32. How does a 'sclerobasis' differ from a true corallum?
33. Compare the nutrient system in
    a. Actinia and Pleurobrachia.
    b. Pleurobrachia and Beroe.
34. Describe the structural relations of the tentacles in
    a. Actinia;
    b. Cestum;
    c. Pleurobrachia.
35. What characters distinguish the thread-cells of the Ctenophora?
36. Describe the structure and position of the 'ctenocyst.'
37. Give some account of the nervous system of the Ctenophora.
38. What peculiarity of position distinguishes the reproductive organs of Tubipora?
39. Define precisely the position of the male and female organs in the Ctenophora.
40. What, according to Haime, is the number and succession of the tentacles in the common Sea-anemone?
41. Give some account of the development of the canal system in Beroe.
42. What numerical law governs the development of the 'septa' in a Zoantharian corallite?
43. Distinguish three principal modes of gemmation among the coralligenous Actinozoa.
44. Describe the structure of a Fringing-reef.
45. How has Mr. Darwin explained the true nature of Barrier-reefs and Atolls?

46. Define the characters of the family Beroidæ, with reference to the subjoined categories:
    - *a.* form of body;
    - *b.* mouth;
    - *c.* canal system;
    - *d.* tentacles.
47. Give some account of the distribution, bathymetrical and geographical, of the reef-building Corals.
48. What families of Zoantharia seem wholly extinct?
49. In what deposits does the family of Astræidæ first make its appearance?
50. Name those groups of Corals which are most abundantly represented in the Paleozoic series.

LIST OF ILLUSTRATIONS.

	Page
1. Urticating organs of *Cœlenterata*, after Gosse	4
2. Development of *Cœlenterata*	15
3. Morphology of *Hydra*, after Hancock, Johnston, and Allen Thomson.	24
4. Morphology of *Hydrozoa*	26
5. Morphology of *Cordylophora*, after Allman	30
6. Reproductive processes of *Hydrozoa*, after Gegenbaur	42
7. Oceanic forms of *Lucernariæ*, after Gosse.	48
8. Development of *Cordylophora*, after Allman	53
9. Development of *Tubularia indivisa*, after Mummery	55
10. Development of *Campanularia*, after Loven	56
11. Development of *Physalia*, after Huxley	58
12. Development of *Lizzia*, after Claparède	60
13. Development of *Turris*, after Gosse	63
14. Gemmation of *Medusoids*, after Forbes and the Author	64
15. Development of *Chrysaora*, after Dalyell	67
16. Morphology of *Tubulariadæ*, after Alder, Forbes, and Stretthill Wright	84
17. Various forms of *Coryniadæ*, after Alder, Forbes, and Sars	86
18. Morphology of *Sertulariadæ*, after Alder, Dalyell, Forbes, and Johnston	91
19. Morphology of *Campanulariadæ*, after Alder and Hincks	94
20. Morphology of *Calycophoridæ*, after Kölliker	97
21. Morphology of *Velella*, after Kölliker	104
22. Morphology of *Physophoridæ*, after Kölliker	109
23. Morphology of *Medusidæ*	116
24. Various forms of *Medusidæ*.	117
25. *Lucernaria*, after Johnston.	122

		Page
26.	Morphology of *Actinozoa*, after Gosse and Hollard	133
27.	Morphology of *Pleurobrachia*, altered from Agassiz and Huxley	145
28.	Morphology of *Zoantharia Sclerodermata*, after Huxley	156
29.	*Columnaria Franklinii*	158
30.	*Tubipora musica*	159
31.	Development of *Pleurobrachia*, after Strethill Wright	181
32.	*Turbinaria palifera*, after Dana	188
33.	*Dendrophyllia nigrescens*, after Dana	189
34.	*Zoantharia Malacodermata*, after Couch, Forbes, and Gosse	197
35.	*Zoantharia Sclerobasica*, after Dana	202
36.	*Zoantharia Sclerodermata*, after Milne Edwards and Haime	204
37.	*Pennatulidæ* and *Gorgonidæ*, after Patterson, Milne Edwards and Haime	211
38.	*Zaphrentis cylindrica*	215
39.	Various forms of *Ctenophora*, after De Blainville, Gegenbaur, Lesson, Patterson, and the Author.	217

INDEX.

Abyla, 98, 99, 101.
Acalèphes Simples, 114.
Acalèphes Hydrostatiques, 114.
Acanthocœnia, 246.
Acanthocyathus, 240.
Acaulis, 87, 88, 89.
Acervularia, 158, 244.
Acontia, 135.
Acraspeda, 120.
Acrocyst, 95.
Actinacis, 247.
Actinia, 131, 146, 166, 171, 199. 235.
Actinidæ, 205.
Actinoloba, 198.
Actinomeres, 143.
Actinopsis, 198.
Actinosome, 143.
Actinozoa, type of, 131; general characters of, 19, 138; development of, 170; classification of, 196; distribution of, 231; relations of, to time, 236.
Adamsia, 150, 198.
Adelastræa, 239.
Ægina, 116.
Æginidæ, 120.
Æginopsis, 60, 116.
Æquoridæ, 120.
Agalma, 102, 108, 110, 111, 112.
Agamogenesis, 74.
Air-vesicle, of *Physophoridæ*, 101.
Alcyonaria, general characters of, 139, 196, 208; nutritive cavity of, 141; tentacles of, 148; thread-cells of, 151; corallum of, 154, 159, 162; development of, 170; families of, 213; distribution of, 231, 233; relations of, to time, 237, 243.
Alcyonidæ, 213; corallum of, 160, 162, 209.
Alcyonium, 161, 208, 232, 235.
Alimentary canal, of *Cœlenterata*, 3, 29, 141.
Alternate generation, 76.
Alveolites, 238.
Amœboid condition of *Hydra*, 11. 52.
Amplexus, 238.

Anabacia, 247.
Androphore, 45.
Anemonia, 199.
Angeastræa, 246.
Angel-band, 34.
Anisophyllum, 244.
Anomophyllum, 246.
Antagonism, between development and reproduction, 72.
Antennularia, 57, 127.
Anthea, 182, 199, 201.
Antipathes, 200, 201.
Antipathidæ, 206.
Apertures, of somatic cavity in *Actinozoa*, 142.
Apical area, of *Ctenophora*, 143, 145, 219.
Apical canals, 145.
Apical pores, 142, 145.
Apolemia, 102, 108, 110, 111.
Apolemiadæ, 113.
Aporosa, 158, 161, 162, 203, 206, 237, 242.
Appendages, of *Hydrozoa*, 27, 69; of *Hormiphora*, 152; of *Ctenophora*, 218.
Arachnactis, 163, 199.
Aræacis, 247.
Aspidiscus, 246.
Astrangia, 241.
Astrocœnia, 239.
Astræa, 241.
Astræidæ, 203, 207, 232, 237, 245—7.
Astræopora, 241.
Astrohælia, 247.
Athorybia, 107, 108, 110, 112.
Athorybiadæ, 113.
Atolls, 191—4.
Atractylis, 62, 88, 89.
Aulacophyllum, 238.
Aulophyllum, 245.
Aulopora, 204, 244.
Auloporidæ, 204, 206, 237, 244, 245.
Aurelia, 47, 64, 123, 127, 128.
Axophyllum, 245.
Axopora, 247.
Axosmilia, 246.

Balanophyllia, 241.
Barrier-reefs, 191—4.
Baryhælia, 246.
Baryphyllum, 244.
Barysmilia, 240.
Basal gemmation, of *Actinozoa*, 182, 183.
Basal membrane, of *Adamsia*, 150, 198.
Base, of *Actinia* and its allies, 131, 198.
Bathycyathus, 240.
Bathymetrical distribution, of *Hydrozoa*, 126; of *Actinozoa*, 231.
Battersbyia, 244.
Beaumontia, 238.
Beroe, tactile organs of, 168; development of, 180; size of, 216; mouth of, 218; canals of, 220-1.
Beroidæ, form of body in, 216; lappets of, 218; apical filaments of, 219; canal system of, 220—2; definition of, 231.
Bimeria, 85, 87, 88.
Blastoderm, 16.
Blastotrochus, 202.
Body-layers, of *Cœlenterata*, 3, 10, 17.
Body-substance, of *Protozoa*, 7; of *Cœlenterata*, 10.
Body-wall, of *Hydrozoa*, 25; of *Actinozoa*, 136, 149.
Bolina, development of, 179; size of, 216; lobes and earlets of, 218; canals of, 220—3; tentacles of, 227; phosphorescence of, 230.
Bougainvillea, 62.
Brachycyathus, 246.
Brachyphyllia, 240.
Brain-stone Coral, 187.
Briareum, 232.

Calamophyllia, 239.
Calice, 155.
Calicular gemmation, of *Actinozoa*, 183, 184.
Callianira, 217, 223, 227, 234.
Callianiridæ, 230.
Calycophoridæ, general characters of, 80, 96; development of, 57; somatocyst of, 96; cœnosarc of, 98; nectocalyces of, 98; polypites of, 29; tentacles of, 32—4; hydrophyllia of, 99; gonophores of, 99; families of, 100; distribution of, 128.
Calymma, 234.
Calymmidæ, 230.
Campanularia, 44, 47, 57, 90—5, 126.
Campanulariadæ, 94, 95.
Campanulina, 93.
Campophyllum, 238.
Canal system, of *Medusidæ* and *Lucernaridæ*, 48, 115, 121—4; of *Ctenophora*, 144—7, 219—24.
Capnea, 199.

Carboniferous Corals, 244.
Carduella, 121.
Caryophyllia, 240.
Cassiopeia, 49, 67.
Cell-theory, 12.
Centrocœnia, 240.
Cephea, 49, 67, 123.
Ceratotrochus, 247.
Cerianthus, 142, 150, 162, 172, 200, 201.
Cestidæ, 230.
Cestum, form and size of, 216; canals of, 223; tentacles of, 227; secretive organ of, 147; phosphorescence of, 230; distribution of, 234.
Chambered cnidæ, 150.
Charybdea, 126.
Chiaja, papillæ of, 149; development of, 180; canals of, 220—3; tentacles of, 227.
Chætetes, 238, 241, 245.
Chonaxis, 245.
Chonophyllum, 244.
Chonostegites, 244.
Chrysaora, 47, 64, 123, 125, 127.
Cilia, of *Cœlenterata*, 3; of *Ctenophora*, 165.
Cinclides, of *Actinia* and its allies, 135, 142.
Circophyllia, 247.
Circulation, in *Cœlenterata*, 6, 31, 142, 147, 209.
Cladangia, 247.
Cladocora, 240.
Cladophyllia, 239.
Classification, of *Cœlenterata*, 14, 19; of *Hydrozoa*, 79; of *Actinozoa*, 196; of *Corynidæ*, 89; of *Sertularidæ*, 95; of *Calycophoridæ*, 100; of *Physophoridæ*, 113; of *Medusidæ*, 119; of *Lucernaridæ*, 125; of *Zoantharia*, 205; of *Alcyonaria*, 213; of *Rugosa*, 215; of *Ctenophora*, 230.
Clara, 41, 46, 82, 86, 88.
Clavatella. 87, 88, 89.
Clisiophyllum, 238.
Cnidæ, 3, 150.
Cœlenterata, general characters of, 3, 6, 13, 19; classes of, 14, 19; thread-cells of, 3; minute structure of, 10; development of, 14—18.
Cœlosmilia, 239.
Cœnenchyma, 155, 157.
Cœnites, 244.
Cœnocyathus, 202.
Cœnosarc, 27; of *Corynidæ*, 82—3; of *Sertularidæ*, 90; of *Calycophoridæ*, 98; of *Physophoridæ*, 101—3, 107; of *Actinozoa*, 139, 157, 201, 208.
Cæspitose corallum, 186.
Columella, 155.
Column, of *Zoantharia*, 198, 199.
Columnaria, 244.
Combophyllum, 244.
Combs, of *Ctenophora*, 165.

INDEX. 265

Comoseris, 246.
Conocyathus, 247.
Constellaria, 244.
Continuous development, 73.
Conversion, chemical, of tissues, 9.
Coppinia, 91.
Coral, 153.
Coral of commerce, 234.
Coral-reefs, 191—5.
Corals, 140, 153.
Corallite, 155.
Corallium, pores of, 142; corallum of, 154, 210; distribution of, 232, 234.
Corallum, structure of, 153—62; development of, 173—8, 185—90; of *Zoantharia*, 201—5; of *Alcyonaria*, 159, 209; of *Rugosa*, 214.
Corbula, of *Plumularia cristata*, 95.
Cordylophora, gonophores of, 44, 45, 88; development of, 54; cœnosarc of, 82; tentacles of, 86; distribution of, 126, 127.
Cornularia, 160, 209.
Coronets, 155.
Corynactis, 200, 201.
Coryne, gonophores of, 41, 88, 89; tentacles of, 86; distribution of, 126.
Coryniadæ, 89.
Corynitæ, general characters of, 79, 82; tentacles of, 32, 85—8; cœnosarc of, 82; polypary of, 35, 85; polypites of, 29, 85; gonophores of, 40—6, 88, 89; development of, 54; families of, 89; distribution of, 126, 127.
Costæ, 156.
Craspeda, 134.
Craspedota, 120.
Cretaceous Corals, 246.
Cryptangia, 247.
Cryptocarpæ, 120.
Cryptolaria, 128.
Ctenocyst, 144, 166.
Ctenophora, type of, 142; general characters of, 139, 142, 196, 216; form and size of, 140, 216—9; canal system of, 144, 219; tentacles of, 225; tegumentary organs of, 149, 152; thread-cells of, 151; ctenophores of, 164; nervous system of, 167; organs of sense of, 166, 168; reproductive organs of, 169; habits of, 229; phosphorescence of, 230; development of, 178; families of, 230; distribution of, 233—5.
Ctenophoral canals, 146, 219.
Ctenophores, 143, 164.
Cunina, development of, 60, 61; pouches of, 116.
Cyanea, 47, 64, 127.
Cyathaxonia, 214, 238.
Cyathaxonidæ, 216, 237.
Cyathophora, 239.
Cyathophyllidæ, 216, 237, 244, 245.

Cyathophyllum, 184, 215, 238.
Cyathoseris, 247.
Cycle, of septa, 175.
Cyclocyathus, 246.
Cyclolites, 240.
Cycloseris, 240.
Cyclosmilia, 247.
Cydippe, 142.
Cyphastræa, 240.
Cystiphyllidæ, 216, 237, 244.
Cystiphyllum, 215, 238.

Dactylosmilia, 246.
Danaia, 244.
Dasmia, 247.
Dasmidæ, 207, 237, 247.
Dasyphyllia, 240.
Definite gemmation, 184.
Dekaia, 244.
Deltocyathus, 247.
Dendracis, 247.
Dendrophyllia, 241.
Dendropora, 244.
Dendrosmilia, 247.
Dendrostyles, of *Rhizostomidæ*, 49.
Depastrum, 121.
Deposition, chemical, in tissues, 9.
Dermosclerites, 160.
Desmophyllum, 240.
Development, of animals, 14, 18, 70; of *Cœlenterata*, 14, 17, 18; of *Hydrozoa*, 51; of *Actinozoa*, 170; of *Hydridæ*, 51; of *Corynidæ*, 54; of *Sertularidæ*, 56; of *Calycophoridæ*, 57; of *Physophoridæ*, 57; of *Medusidæ*, 60; of *Lucernaridæ*, 64; of *Zoantharia* and *Alcyonaria*, 170, 185; of *Ctenophora*, 178.
Devonian Corals, 244.
Deuterozoöids, 74.
Dicoryne, 88.
Digestive sac, of *Actinozoa*, 132, 141, 144, 220.
Dimorphastræa, 247.
Diphydæ, 100.
Diphyes, 50, 96, 99, 100, 128.
Diphyozoöids, 100, 127.
Diploctenium, 246.
Diplohælia, 247.
Diplonema, 63.
Diploria, 240.
Disc, gelatinous, of *Medusidæ* and *Lucernaridæ*, 34; of *Actinia* and its allies, 131, 198.
Discocyathus, 246.
Discontinuous development, 73.
Discosoma, 199.
Discotrochus, 247.
Dissepiments, 155.
Distal extremity, of hydrosoma, 25; of polypite, 29.
Distal nectocalyx, of *Diphyes* and *Abyla*, 96, 98.
Distribution, of *Cœlenterata*, 13; of *Hydrozoa*, 126; of *Actinozoa*, 231.

Earlets, of *Ctenophora*, 218.
Ecderon, 149.
Echinoporidæ, 207, 237.
Ecmesus, 247.
Æcthoræum, 4.
Ectoderm, 3, 10, 17.
Edwardsia, 162, 199.
Elasmocœnia, 246.
Embryo, of *Hydrozoa*, 51; of *Actinozoa*, 170.
Emmonsia, 238.
Enallohœlia, 239.
Enderon, 149.
Endoderm, 3, 10, 17.
Endopachys, 241.
Endophyllum, 244.
Endoplasts, 10.
Ephyra, 66
Epitheca, 157.
Epithelial layer, of disc in *Medusidæ* and *Lucernaridæ*, 35; of body-wall in *Actinozoa*, 136, 149.
Eridophyllum, 238.
Eucopidæ, 120.
Eudendrium, 83, 87—9.
Euhælia, 246.
Eumenides, 148, 200.
Euphyllia, 240.
Eupsammia, 203, 241.
Eurhamphæa, 218, 220, 223, 227, 228.
Eurymœandra, 239.
Eurystomata, 218, 231.
Exothecæ, 156.
Eye-specks, 38, 166.

Families, of *Corynidæ*, 89; of *Sertularidæ*, 95; of *Calycophoridæ*, 100; of *Physophoridæ*, 113; of *Medusidæ*, 119; of *Lucernaridæ*, 125; of *Zoantharia*, 205; of *Alcyonaria*, 213; of *Rugosa*, 215; of *Ctenophora*, 230.
Fan-Corals, 186.
Favia, 239.
Favosites, 238.
Favositidæ, 205, 206, 237, 244, 245.
Fecundation, in *Hydrozoa*, 50; in *Actinozoa*, 170.
Feelers, of *Physophoridæ*, 40.
Fibrillation, of tissues, 8.
Filament, of tentacle in *Hydrozoa*, 33.
Firm part, of *Velella*, 104.
Fission, 72; of *Hydra*, 52; of Medusoids, 63; of *Stomobrachium*, 63; of Hydra-tuba, 66; of *Actinozoa*, 182, 185.
Fistulipora, 238.
Fletscheria, 244.
Float, of *Physophoridæ*, 27, 101.
Foot secretion, 153.
Form of body, in *Cœlenterata*, 6; in *Hydrozoa*, 25—7; in *Actinozoa*, 196, 208, 216.
Formless contractile substance, 52.
Forskalia, 34, 102, 108, 111.
Fossil Corals, 236.

Fossil forms, of *Hydrozoa*, 130; of *Actinozoa*, 236—47.
Fresh-water forms of *Cœlenterata*, 13, 126.
Fringing-reefs, 191—4.
Fungia, 199.
Fungidæ, 203, 207, 232, 237, 244, 246, 247.
Funnel, of *Ctenophora*, 144, 220.

Galaxea, 240.
Gamogenesis, 75.
Ganglia, of *Actinia*, 166; of *Ctenophora*, 167.
Garveia, 88.
Gastric filaments, of *Lucernaridæ*, 124.
Gastric region, of polypite, 29.
Gemmation, 72; of *Cœlenterata*, 6; of *Hydrozoa*, 51, 54, 57, 63, 65, 69; of *Actinozoa*, 182—90.
Genabacia, 247.
General cavity, of *Hydrozoa*, 17, 29; of *Actinozoa*, 17, 141.
General morphology, of *Cœlenterata*, 1, 6, 10; of *Hydrozoa*, 25; of *Actinozoa*, 138.
Geographical distribution, of *Hydrozoa*, 127; of *Actinozoa*, 233.
Germinal dot, 15.
Germinal vesicle, 15.
Geryonia, marginal vesicle of, 38.
Geryonidæ, 120.
Glandular sacs, of *Velella*, 106.
Globate cnidæ, 150.
Goniastræa, 240.
Goniocora, 239.
Goniophyllum, 244.
Gonoblastidia, 45; of *Corynidæ*, 46, 88; of *Sertularidæ*, 46, 94; of *Physophoridæ*, 112.
Gonocalyx, 43.
Gonophores, structure of, 40—5; position of, 45—7; of *Corynidæ*, 41, 44, 88; of *Sertularidæ*, 44, 46, 93; of *Calycophoridæ*, 45, 99; of *Physophoridæ*, 45, 111.
Gonothecæ, 47.
Gorgonia, 161, 232, 235.
Gorgonidæ, general characters of, 163, 210, 214; somatic cavity of, 141; sclerobasis of, 154, 162; spicules of, 161; development of, 185; distribution of, 232; relations of, to time, 237, 246.
Graphularia, 247.
Graptolites, 130.
Gymnophthalmata, 120.
Gynophore, 45.

Hadrophyllum, 244.
Haimeia, 208.
Halecium, 93.
Haticornaria, 93.
Halistemma, 103, 108, 111, 112.

Hallia, 244.
Halysites, 244.
Haplocœnia, 247.
Haplophyllia, 239.
Haplosmilia, 246.
Heliastræa, 239.
Heliolites, 238.
Heliophyllum, 244.
Heteractis, 200.
Heterocœnia, 246.
Hippopodidæ, 100.
Hippopodius, 45, 98, 99.
History, of *Hydrozoa*, 130; of *Actinozoa*, 236; of *Zoantharia*, 242; of *Rugosa*, 243; of *Alcyonaria*, 243.
Hœmal region, 17.
Holocœnia, 246.
Holocystis, 214, 247.
Hormiphora, 152.
Hyalochætidæ, 201, 206.
Hyalonema, 154.
Hyalopathes, 154, 241.
Hybocœnia, 239.
Hydnophora, 240.
Hydra, morphology and physiology of 20—5; development of, 51; amœboid condition of, 11, 52; species of, 82.; distribution of, 126, 127.
Hydra-tuba, 65—7, 124.
Hydractinia, tentacles of, 32, 87; cœnosarc of, 83; polypary of, 85; gonophores of, 41, 45; gonoblastidia of, 46.
Hydridæ, 51, 79, 80.
Hydrocysts, of *Physophoridæ*, 40, 110.
Hydrœcium, of *Calycophoridæ*, 96, 99.
Hydrophyllia, 35; of *Calycophoridæ*, 99; of *Physophoridæ*, 108.
Hydrorhiza, 26.
Hydrosoma, 25.
Hydrothecæ, of *Sertularidæ*, 27, 70, 92.
Hydrozoa, type of, 20; general characters of, 19, 20; development of, 51; classification of, 79; distribution of, 126.; relations of, to time, 130.
Hymenophyllia, 246.

Ilyanthus, 198.
Indefinite gemmation, 184.
Individuality, of animals, 13, 71.
Inner layer, of blastoderm, 16.
Integument, of *Cœlenterata*, 10; of *Hydrozoa*, 34; of *Actinozoa*, 149.
Intermediate layer, 16.
Involucrum, 34.
Irregular gemmation, 184.
Isastræa, 239.
Isis, 154, 210, 240, 241.

Jelly-fishes, 13, 127.
Jurassic Corals, 245.

Koninckia, 247.
Kophobelemnon, 212.

Labechcia, 244.
Lagoon, 192.
Lamellar corallum, 187.
Lappets, of *Beroidæ*, 218.
Lar, 88.
Layers, of *Cœlenterata*, 3, 10.
Leptocyathus, 247.
Leptophyllia, 246.
Leptoria, 240.
Lesueuria, canals of, 222, 223; tentacles of, 228.
Leucothea, tentacles of, 228.
Life, duration of, in *Actinozoa*, 182.
Life-history, of *Hydrozoa*, 51; of *Actinozoa*, 170.
Limb, of *Velella*, 104.
Lincolaria, gonophores of, 128.
Liriope, 60, 115.
Litharæa, 247.
Lithocysts, of *Lucernaridæ*, 38, 124.
Lithophyllia, 240.
Lithostrotion, 238.
Liver, of *Hydrozoa*, 31, 106; of *Actinozoa*, 137, 147.
Lizzia, development of, 60, 63.
Lobopsammia, 247.
Locomotive organs, of *Hydrozoa*, 36; of *Actinozoa*, 164.
Loculi, 155.
Lonsdaleia, 245.
Lophohœlia, 240.
Lophophyllum, 238.
Lophosmilia, 240.
Lucernaria, 40, 121, 124, 125.
Lucernariadæ, 67, 120, 122—5.
Lucernaridæ, general characters of, 80, 120—5; umbrella of, 27, 37, 124; lithocysts of, 38, 124; canal system of, 48, 124; gastric filaments of, 124; reproductive zoöids of, 47; development of, 64, 123; families of, 125; distribution of, 127, 128.
Lucernaroids, 123.
Lyellia, 244.

Madrepora, 241.
Madreporidæ, 203, 207, 232—4, 237, 246, 247.
Malacodermata, 201, 205, 237.
Manubrium, 43.
Marginal bodies, of *Medusidæ*, 37; of *Lucernaridæ*, 38, 124; of *Actinia*, 165.
Massive corallum, 187.
Meconidia, of *Campanularia Loveni*, 95.
Medusidæ, general characters of, 80, 114—20; nectocalyx of, 36, 116; marginal bodies of, 37; polypite of, 115; tentacles of, 117; reproductive organs of, 118; development of, 60, 118; families of, 119; distribution of, 128; phosphorescence of, 129.
Medusiform gonophores, 45.
Medusoids, 43, 45, 62, 118.

Melitæa, 154, 210.
Membrana intermedia, 16.
Menophyllum, 245.
Mertensia, 235.
Merulina, 208.
Merulinidæ, 208, 237.
Mesenteries, of *Actinozoa*, 132—4, 137, 146, 172.
Metagenesis, 74.
Metamorphosis, 73.
Metastræa, 247.
Michelinia, 238.
Micrabacia, 247.
Micropyle, of animal ovum, 71.
Microsolena, 246.
Millepora, 239.
Milleporidæ, 205, 206 237, 244, 247.
Minyas, 163.
Mæandrastræa, 246.
Mæandrina, 142, 187, 199, 203, 232, 239.
Monœcious forms, of *Hydrozoa*, 50, 112; of *Actinozoa*, 169, 229.
Montlivaultia, 239.
Mopsea, 154, 210, 241.
Mouth, of *Hydrozoa*, 29, 86, 88; of *Actinozoa*, 131, 141; of *Ctenophora*, 218.
Movements, of *Hydrozoa*, 23, 27, 36; of *Actinozoa*, 138, 162, 210; of *Ctenophora*, 229.
Mucous layer, of blastoderm, 16.
Mulberry-mass, 16.
Muscular system, of *Hydrozoa*, 36; of *Actinozoa*, 162.
Mycetophyllia, 240.
Myriothela, 82, 87.

Nectocalyces, 27, 36, 70; of *Calycophoridæ*, 96—9; of *Physophoridæ*, 107; of *Medusidæ*, 116.
Nectocalycine canals, 37.
Nectosac, 36.
Nematophores, of *Sertularidæ*, 34, 93.
Nemopsis, 84.
Nervous system, of *Medusidæ*, 39; of *Actinia*, 166; of *Ctenophora*, 167.
Neural region, 17.
Nutritive organs, of *Hydrozoa*, 28; of *Actinozoa*, 141.

Occania turrita, marginal bodies of, 38.
Oceanidæ, 119.
Ocelli, 38.
Oculina, 240.
Oculinidæ, 203, 207, 237, 246, 247.
Ocyroe, 234.
Oldhamiæ, 130.
Omphyma, 244.
Operculum, of *Campanularia fastigiata*, 92.
Oral canals, of *Ctenophora*, 220.
Order, of septa, 175.
Orders, of *Hydrozoa*, 79; of *Actinozoa*, 196.

Organ-pipe Corals, 209.
Oroseris, 239.
Otolites, 39.
Outer layer, of blastoderm, 16.
Ova, of animals, 14, 71; of *Hydrozoa*, 50; of *Actinozoa*, 169.
Ovaria, of *Hydrozoa*, 40, 50; of *Actinozoa*, 169.
Ovigerous vesicles, of *Sertularidæ*, 47.

Pachygyra, 239.
Pachyphyllum, 244.
Pali, 155.
Palæocyclus, 244.
Palpocils, 93.
Palythoa, 141.
Paracyathus, 240.
Paragastric canals, of *Ctenophora*, 145, 220.
Parasitic forms, of *Hydrozoa*, 61, 83; of *Actinozoa*, 198, 233.
Parasmilia, 246.
Parietal gemmation, of *Actinozoa*, 183.
Parthenogenesis, 74; theory of, 78.
Paronaria, 240.
Peachia, 142, 199, 200.
Pedicle, of tentacle in *Hydrozoa*, 33.
Peduncle, of polypite, 29, 31.
Pelagia, structure of, 40, 123; development of, 68; phosphorescence of, 129.
Pelagidæ, 125.
Pennaria, 87—9.
Pennatula, 212—3, 232.
Pennatulidæ, 163, 186, 210—3, 214, 235, 237, 246, 247.
Pentacœnia, 246.
Peplosmilia, 246.
Perforata, 158, 161, 203, 206, 237, 242.
Perigonimus, 87 9.
Periplast, 10.
Peristomial space, 132, 198.
Permian Corals, 245.
Phanerocarpæ, 120.
Phillipsastræa, 238.
Philomedusa, 142, 233.
Phosphorescence, of *Cœlenterata*, 7; of *Hydrozoa*, 128; of *Pennatulidæ*, 212; of *Ctenophora*, 230.
Phyllactis, 200.
Phyllangia, 240.
Phyllocœnia, 240.
Phyllocyst, 36.
Phyogemmaria, of *Velella*, 105.
Physalia, thread-cells of, 5; villi of, 31; tentacles of, 32, 33, 111; development of, 58; pneumatophore of, 101, 107; gonophores of, 111, 112; distribution of, 127.
Physaliadæ, 114.
Physical elements, relations to, of *Hydrozoa*, 126; of *Actinozoa*, 231.
Physophora, 102, 107—12.
Physophoriadæ, 113.
Physophoridæ, general characters of,

INDEX. 269

80, 101; development of, 57; pneumatophore of, 101, 107; cœnosarc of, 107; nectocalyces of, 107; hydrophyllia of, 108; polypites of, 29, 110; hydrocysts of, 40, 110; tentacles of, 32—4, 111; gonophores of, 47, 111; families of, 113; distribution of, 127, 128.
Phytogyra, 246.
Pigment masses, of *Hydrozoa*, 36, 38; of *Actinozoa*, 152.
Placocœnia, 246.
Placosmilia, 246.
Placosphyllia, 246.
Plan of structure, of *Cœlenterata*, 3, 19; of *Hydrozoa*, 17, 19, 25; of *Actinozoa*, 17, 19, 138.
Planula, 64.
Platytrochus, 247.
Plerastræa, 239.
Plesiastræa, 241.
Pleurobrachia, form and structure of, 143, 216, 221; thread-cells of, 151; nervous system of, 168; tactile organs of, 168; development of, 180; tentacles of, 225; distribution of, 235.
Pleurobrachiadæ, 230.
Pleurocœnia, 239.
Pleurocora, 247.
Pleurodictyum, 244.
Plumularia, 34, 44, 50, 93, 94.
Pneumatic filaments, of *Velella* and *Porpita*, 102, 106.
Pneumatocyst, of *Physophoridæ*, 101; of *Velella*, 105.
Pneumatophore, of *Physophoridæ*, 101, 107.
Podactinaria, 120.
Podocoryne, 88, 112.
Polycœlia, 245.
Polypary, of *Corynidæ* and *Sertularidæ*, 35, 85, 92.
Polype, 139; of *Zoantharia*, 197; of *Alcyonaria*, 208.
Polype-cells, of *Sertularidæ*, 35.
Polypite, 25, 29; of *Hydridæ*, 20, 82; of *Corynidæ*, 85; of *Sertularidæ*, 92; of *Calycophoridæ*, 29, 99; of *Physophoridæ*, 110; of *Medusidæ*, 115; of *Lucernaridæ*, 47, 65, 121, 123.
Polytremacis, 247.
Porites, 232, 241.
Poritidæ, 204, 207, 237, 244, 246, 247.
Porpita, 32, 45, 102, 103, 107, 110—112.
Pouches, of *Æginidæ*, 116, 120.
Praya, 98, 99.
Prayidæ, 100.
Prehensile organs, of *Hydrozoa*, 32; of *Actinozoa*, 148.
Prionastræa, 241.
Propora, 238.
Protaræa, 244.
Protoplasm, animal, 7.

Protoseris, 246.
Protovirgularia, 243.
Protozoa, their minute structure, compared with that of *Cœlenterata*, 7; development of, 14, 18.
Protozoöids, 74.
Proximal extremity, of hydrosoma, 25; of polypite, 29.
Psammopora, 238.
Pterygia, 150.
Ptychophyllum, 238.
Pyloric valve, of *Calycophoridæ*, 29.
Pyrgia, 204, 245.

Radial canal system, of *Ctenophora*, 219, 221.
Recent *Actinozoa*, 248.
Regular gemmation, 184.
Reniform enlargements, of tentacle in *Physalia*, 33, 111.
Renilla, 212.
Reparative powers, of *Hydrozoa*, 52, 55; of *Actinozoa*, 182.
Reproductive organs, of *Hydrozoa*, 40; of *Actinozoa*, 134, 169; of *Hydridæ*, 23; of *Corynidæ*, 88; of *Sertularidæ*, 93; of *Calycophoridæ*, 99; of *Physophoridæ*, 111; of *Medusidæ*, 40, 118; of *Lucernaridæ*, 40, 47; of *Zoantharia* and *Alcyonaria*, 134, 169, 209; of *Ctenophora*, 169.
Respiration, in *Cœlenterata*, 6, 209.
Reticularia, 91.
Rhabdophyllia, 239.
Rhabdopora, 245.
Rhipidogyra, 239.
Rhizangia, 240.
Rhizophysa, 101—2, 107, 110—1.
Rhizophysiadæ, 114.
Rhizostoma, 49, 67.
Rhizostomidæ, 49, 125.
Rhodaræa, 241.
Rœmeria, 244.
Rugosa, corallum of, 158, 173, 214; families of, 215; relations of, to time, 237, 243.

Saccanthus, 199, 200.
Sacculus, of tentacle in *Hydrozoa*, 33.
Sarcode, 7, 11.
Sarcodictyon, 151, 160, 209.
Sarsia, gemmation of, 63.
Sclerenchyma, 161.
Sclerites, of *Gorgonidæ*, 161.
Sclerobasic corallum, 153; development of, 185.
Sclerobasic Zoantharia, 154, 162, 201, 234, 237.
Sclerodermic corallum, 155, 173, 186.
Sclerodermic Zoantharia, 158, 161, 173, 202, 234, 237.
Scyphistoma, 66.
Sea-Anemones, 131, 162, 198.
Sea-Fir, 90.

Sea-Pens, 163, 212.
Sea-Shrubs, 154.
Secretive organ, of *Cestum*, 147.
Segmentation, of ovum, 15
Sense-organs, of *Hydrozoa*, 39; of *Actinozoa*, 166, 168.
Septa, 155; development of, 173.
Septal formulæ, 174.
Seriatoporidæ, 205, 206, 232, 237, 244, 245.
Serous layer, of blastoderm, 16.
Sertularia, 90, 95, 127.
Sertulariadæ, 77, 92, 95.
Sertularidæ, general characters of, 79, 90; development of, 56; cœnosarc of, 90; polypary of, 35, 92; hydrothecæ of, 35, 92; polypites of, 92; tentacles of, 93; nematophores of, 93; gonoblastidia of, 46; gonophores of, 44, 93; families of, 95; distribution of, 126—8.
Silurian Corals, 243.
Size, of *Cœlenterata*, 6; of *Hydrozoa*, 28; of *Actinozoa*, 140.
Smilotrochus, 246.
Smithia, 244.
Sidenastræa, 241.
Somatic cavity, of *Cœlenterata*, 3, 10, 17; of *Hydrozoa*, 17, 29, 31; of *Actinozoa*, 17, 141, 144—7.
Somatic chambers, of *Actinozoa*, 134, 146.
Somatic fluid, 6.
Somatocyst, of *Calycophoridæ*, 96.
Spermaria, of *Hydrozoa*, 40, 50; of *Actinozoa*, 169.
Spermatozoa, 15; of *Hydrozoa*, 50; of *Actinozoa*, 169
Sphenotrochus, 246.
Sphæronectes, 98, 99.
Sphæronectidæ, 100.
Spicular corallum, 160, 162, 209, 212.
Spicules, of *Alcyonidæ* and *Zoanthidæ*, 160; of *Gorgonidæ*, 161.
Spiral cnidæ, 150, 151.
Spongophyllum, 244
Sporosac, 40, 45.
Spurious cœnenchyma, 205.
Stauria, 214, 244.
Stauridæ, 215, 237, 244, 245.
Stauridia, tentacles of, 87; gonophores of, 88, 89.
Steenstrupia, 63.
Steganophthalmata, 120.
Stelloria, 246.
Stenostomata, 219, 230
Stephanocœnia, 239.
Stephanomia, 108, 110, 112.
Stephanomiadæ, 113.
Stephanophyllia, 240.
Stereopsammia, 247.
Stinging powers, of *Cœlenterata*, 5.
Stomach, of *Actinozoa*, 132, 141, 144
Stomachal filaments, of *Lucernaridæ*, 124.

Stomatodendra, of *Rhizostomidæ*, 49.
Stomobrachium, fission of, 63.
Strebla, 151
Streptelasma, 244.
Strobila, 66.
Strombodes, 244.
Stylaxis, 245.
Stylina, 239, 245.
Stylocœnia, 240.
Stylocyathus, 246.
Stylophora, 174, 240.
Stylophoridæ, 207, 237, 247.
Stylosmilia, 239.
Successive development, of appendages in *Hydrozoa*, 70; of mesenteries and tentacles in *Actinozoa*, 132, 172; of septa in corallite, 173.
Swimming-bells, of *Hydrozoa*, 27, 36; of *Calycophoridæ*, 96—9; of *Physophoridæ*, 107; of *Medusidæ*, 115—6.
Symphyllia, 241.
Synapticulæ, 155, 203.
Syndendrium, of *Rhizostomidæ*, 49.
Synhœlia, 246.
Syringophyllum, 238.
Syringopora, 238.

Tabulata, 158, 161, 205, 237, 242.
Tabulæ, 156.
Tactile organs, of *Hydrozoa*, 40; of *Ctenophora*, 168.
Tamoya, 126.
Tangled cnidæ, 150, 151.
Tegumentary organs, of *Hydrozoa*, 34; of *Actinozoa*, 149.
Telestho, 159, 209.
Tentacles, of *Hydrozoa*, 32; of *Actinozoa*, 148; of *Hydridæ*, 20; of *Corynidæ*, 32, 85; of *Sertularidæ*, 32, 93; of *Calycophoridæ*, 32, 33; of *Physophoridæ*, 32, 33, 111; of *Medusidæ*, 32, 117; of *Lucernaridæ*, 32, 121, 125; of *Zoantharia*, 148, 199; of *Alcyonaria*, 148; of *Ctenophora*, 225.
Tertiary Corals, 247.
Thalassianthus, tentacles of, 200.
Thamnastræa, 239.
Thaumantiadæ, 120.
Thaumantias, 63, 129.
Theca, 155.
Thecia, 244.
Thecidæ, 205, 237, 242, 244.
Thecocyathus, 246.
Thecosmilia, 239.
Thecostegites, 244.
Thread-cells, of *Cœlenterata*, 3; of *Physalia*, 5; of *Hydra*, 20; of *Actinozoa*, 150.
Tichastræa, 247.
Time, relations to, of *Hydrozoa*, 130; of *Actinozoa*, 236.
Tissue secretion, 153.
Trachynema, 60, 117.
Trachynemidæ, 119, 120.

INDEX.

Trachypora, 241.
Triassic Corals, 245.
Trichydra, 82, 87, 89.
Tritozoöids, 74.
Trochocyathus, 239.
Trochophyllum, 245.
Trochoseris, 247.
Trochosmilia, 239.
Tubipora, 209.
Tubiporidæ, 159, 162, 209, 213, 237.
Tubularia, 31, 32, 44, 54, 83—9, 127.
Tubulariadæ, 89.
Tubularinæ, 90.
Tubularina, 90.
Tubulosa, 159, 162, 204, 206, 237, 242, 243.
Turbinaria, 241.
Turbinolia, 247.
Turbinolidæ, 202, 207, 233, 234, 237, 246, 247.
Turris, 62.
Turritopsis, 61.
Type, of *Hydrozoa*, 20; of *Actinozoa*, 131.

Ulocyathus, 232.
Ulophyllia, 239.
Umbellularia, 235.
Umbrella, of *Lucernaridæ*, 27, 37, 47, 124.
Urticating organs, of *Cœlenterata*, 3.

Vacuolation, of tissues, 8, 32.
Vegetative repetition, 73.
Veil, of nectocalyx, 36, 116.
Velella, form and structure of, 104;
development of, 59; pneumatocyst of, 102, 105; tentacles of, 32, 111; polypites of, 105, 110, 114; gonophores of, 45, 112—3; glandular sacs of, 106; distribution of, 127.
Velellidæ, 114.
Veretillum, 212.
Vesicles, of *Medusidæ*, 37.
Villi, of polypite, 31.
Virgularia, 212, 241.
Vital endowments, of lower animals, 10.
Vitelline membrane, 15.
Vogtia, 45, 99.
Vorticlava, 82, 85—7, 89.

Warts, of Sea-anemones, 149.
Willsia, canals of, 116.

Yolk, 14.
Yolk-division, 15, 51, 170.
Yolk-sac, 15.

Zaphrentis, 214, 238.
Zoantharia, general characters of, 139, 169, 196; polypes of, 131, 141, 197; tentacles of, 148, 199; threadcells of, 150; corallum of, 158—62, 173, 186, 201—5; free-swimming forms of, 163; families of, 201, 205; development of, 170; distribution of, 232—6; relations of, to time, 237, 242.
Zoanthidæ, 162, 199, 201, 206.
Zoanthus, 160, 235.
Zoöids, 74.

THE END.

LONDON
PRINTED BY SPOTTISWOODE AND CO.
NEW-STREET SQUARE

www.ingramcontent.com/pod-product-compliance
Lightning Source LLC
Chambersburg PA
CBHW032102220426
43664CB00008B/1108